AF553797

# Environmental Analysis and Instrumentation

# Environmental Analysis and Instrumentation

Neelima Rajvaidya
Dilip Kumar Markandey

A.P.H. PUBLISHING CORPORATION
ANSARI ROAD DARYA GANJ
NEW DELHI-110 002

*Published by*
S.B. Nangia
**A P H Publishing Corporation**
4435-36/7, Ansari Road, Daryaganj
New Delhi 110002
Ph.: 23274050
E-mail : aphbooks@gmail.com

2026

Rs. 2995/-



*Printed at*
**Balaji Offset**
Navin Shahdara, Delhi 110032

# CONTENTS

## Chapter -1

# SIMPLE CHEMICAL BALANCE

## INTRODUCTION

Length, mass and time are known as fundamental units, from them, rest of the units are derived. To measure length of a particular object, generally centimeter, meter, kilometer and other silimar forms are in existance. Similarly, time is expressed in the form of seconds, minutes, hours and so on.

To measure the mass of a certain object, extensive range of measurement is also available This measurement scale include following specifications-

1. Milligram
2. Gram
3. Kilogram

For physical measurement of certain object, generally gram/ kilogram units of expression are in use. However for simple chemical analytical measurement, generally milligram and its subsequent dimensions are in practice. In general, following dimensions can be observed, as presented in fig. 1.1.

For physical measurement of mass/weight, double pan balance in use, however, for same measurements, in chemical environment, single pan balance is in operation (Fig. 1.2).

*Description*

Depending upon the nature of assignments, the balance, which is to be used in the concern laboratory, should exhibit the extent of precision and accuracy.

## OPERATIONAL INSTRUCTION OF A TYPICAL CHEMICAL BALANCE

See fig. 1.2 as a model instrument :

1. Set the mark on the knob 'D' at the up position almost in the middle towards the pan.
2. Connect the chord to the mains and switch on. Accordingly, the light will glow in the observation aperture 'B' at the right hand top corner of the instrument (i.e. at your left hand side).
3. Mercury bubble becomes visible in the round aperture 'B'. As the light glow, adjust the mercury bubble in the centre of the ring (aperture) by turning the knob (legs-a & b) either clockwise or anticlockwise, as required.
4. Now, turn the knob 'D' anticlockwise (towards) your left). By doing so, the bulb will glow in the scale frame E and the two scale can be seen there. The scale on your left side is fixed (stationary), whereas the right side scale is movable. Bring 'O' (i.e. zero) of the moving scale against 'O' of the stationary scale by playing the knob 'C', exists on the right side of the balance.
5. If the zeros are not coinciding with the help of this knob 'C', remove the top cover and rotate the fine screw 'S' of the beam, which is provided by balance manufacturing company for preliminary adjustment for operation (not shown in the figure)
6. As the zeros are adjusted, bring the resting point of knob 'D' at the top position.

Now the balance is stabilized and ready for use.

## TO MEASURE THE WEIGHT OF A GIVEN CHEMICAL

1. Now, open the side window and put the material to be weighted on the pan 'F' close the window. Notice that if the liquid chemicals are to be weighted, take the weight of container

separately. Similarly, for weighing the powder chemicals, first place a piece of paper in the pan, take its weight and then take the weight of the chemical, given of weighing.

2. Rotate the knob 'D' a clockwise (to your right hand side) and adjust the approximate weight by the desired knob (s). If it is less than 1 gm., then rotate the knob 'j' (0.1 gm) clockwise, step-by-step (1,2,3 and so on) and observe the movement of the scale. For example, if the moving scale shows movement between 0.3 & 0.4, then fix the knob at 0.3 and then rotate the knob 'D' anticlockwise (to extreme left). The moving scale will become stationary at the some place and its marking will show the reading at the second, third and fourth place of the decimal digit.

Suppose, the moving scale stops at a point, when its scale indicates fifth division (which is coincides) between 20-30 marking against the zero of the fixed scale. This would means that the weight of the given chemical is 0.125 gms.

The second place is represented by 20 of the moving scale and the third place is represented by the division on the moving scale, which coincides with the zero of the fixed scale (i.e. fifth division between 20-30, in this case).

If the weight is to be taken more precisely (means, further upto fourth place), then the reading should be made on the fixed scale at the point, where small division exactly coincides with the small division of the fixed scale. This will be made more clear by the fig 1.3 as a example, in which

The first decimal place-

The first decimal place has already been determined by the knob 'J' at 0.1, the second and third places is 5 because fifth division of moving scale coincides with zero.

The fourth place is 5, the fifth division of the fixed scale coincides with small division of moving in the instrument.

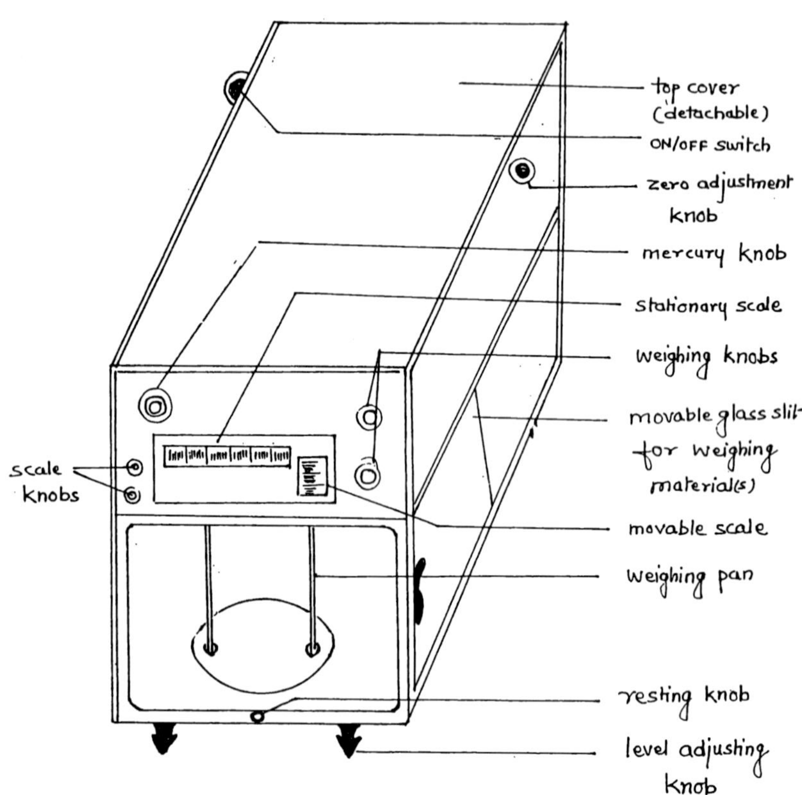


**Fig. 1.1 : Main unit of Chemical balance**

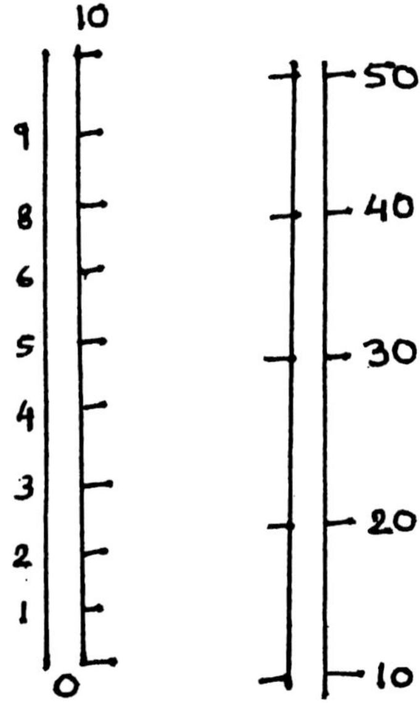


**Fig. 1.2 : Measuring Scale**

Chapter -2

# DIGITAL pH METER

## INTRODUCTION

Concentration of hydrogen ions present in a aqueous solution, is measured in terms of pH.

Physicochemically-

$$pH = \log \frac{1}{H^+}$$

or,

$$= -\log H^+$$

***"The pH is the negative log of the hydrogen ion concentration."***

Thus, mathematically, it can be expressed as -

It is a quantitative expression for acidity of alkalinity of a aqueous solution & ranges from 1 to 14.

An acid may be define as-

***"a substance which, when dissolved in water, produces hydrogen ions ($H^+$)."***

Similarly, a base ***"is a substance, which produces hydroxyl ions ($OH^-$), when dissolved in water."***

A normal solution of any acid will exactly neutralise an equal amount (or volume) of a normal solution of any base, no matter how strong or weak, may be characteristics of two reactants.

During neutralisation, equal numbers of hydrogen ions of an acid and hydroxyl ions of a base, unites to form water molecule. When the number of hydrogen ions is equivalent to the number of hydroxyl ions, water molecules are formed, showing neutral chemical nature and its pH is said to be 7.

At a particular temperature, the product of number of hydrogen ions multiplied by a number of hydroxyl ions, is constant.

This, at 23°C :

$(H^+) \times (OH^-) = 10^{-14}$

For each pH unit less than 7, the concentration of hydrogen ions is increased tenfold. Similarly for each pH unit above 7.0, it is decreased tenfold.

pH 7.0 is considered neutral, less than 7 is acidic and more than 7.0 is alkaline or basic.

Thus, pH is measurement of the extent of acidity or alkalinity of a aqueous solution.

## METHODS OF pH DETERMINATION

The following methods are commonly used for determining pH of an unknown solution-

(1) ***By Litmus paper***

It is a paper impregnated with a organic dye. Blue litmus paper turns red in acidic medium (pH less than 7.0), while red litmus paper turns blue in an alkaline medium (pH higher than 7.0).

(2) ***By pH strips or pH paper***

This paper is available in a special dispenser form (litmus paper booklet), which may be removed for determining the pH of a solution. A portion of this strip is placed in the sample, to be tested and then, compared with colour scale. This can give quantitative measurement and is accurate upto a single pH unit.

(3) ***By universal indicators(s)***

These are the dyes which are used to test the pH of a solution.

As the hydrogen ion concentration of a solution changes rearrangement of this indicator molecules occur, resulting in change in colour. They provide fairly accurate quantitative measurement of pH. The following Table 2.1 gives a list of acid base indicator arranged according to their respective pH limitations.

**Table 2.1 : Acid base indicators**

| *S. No.* | *Name of the indicator* | *Colour indication in various range* | | |
|---|---|---|---|---|
| | | *pH range* | *Acidic environment* | *Alkaline environment* |
| *1* | 2 | *3* | *4* | |
| 1. | Methyl violet | 0.1-1.5 | yellow | Blue |
| 2. | Metacresol purple (acid | 0.5-2.5 | Red | Yellow |
| 3. | Tropeolin | 1.4-2.6 | Red | Yellow |
| 4. | Thymol blue (acid) | 1.2-2.8 | Red | Yellow |
| 5. | Alizarin yellow R (P) | 1.9-3.3 | Red | Yellow |
| 6. | Methyl Yellow | 1.9-4.0 | Red | Yellow |
| 7. | Methyl orange | 3.1-4.4 | Red | Yellow |
| 8. | Bromophenol blue | 3.0-4.6 | Yellow | Blue-violet |
| 9. | Congo red | 3.0-5.2 | Blue | Red |
| 10. | Alizarin sod. Phosphate | 3.7-5.2 | Yellow | Violet |
| 11. | Bromocresol green | 3.8-5.4 | Yellow | Blue |
| 12. | Methyl red | 4.2-6.3 | Red Yellow | |
| 13. | Chlorophenol red | 4.8-6.4 | Yellow | Purple |
| 14. | Bromophenol purple | 5.2-6.8 | Yellow | Purple |
| 15. | Bromophenol red | 5.2-7.0 | Yellow | Red |
| 16. | Bromothymol blue | 6.0-7.6 | Yellow | Blue |
| 17. | Neutral red | 6.0-8.0 | Red | Yellow |

| 1 | 2 | 3 | 4 | |
|---|---|---|---|---|
| 18. | Resolic acid | 6.8-8.2 | Yellow | Red |
| 19. | Azolitmin | 4.5-8.3 | Red | Blue |
| 20. | Phenol red | 6.8-8.4 | Yellow | Red |
| 21. | Neptholpethalein | 7.3-8.7 | Rose | Green |
| 22. | Cresol red | 7.2-8.8 | Yellow | Red |
| 23. | Tropeolin | 7.6-8.9 | Yellow | rose-red |
| 24. | Meta cresol purple (alkaline) | 7.4-9.0 | Yellow | Purple |
| 25. | Thymol blue | 8.0-9.6 | Yellow | Blue |
| 26. | Phenopthalein | 8.3-10.0 | Colourless | Red |
| 27. | Thymopthalein | 9.3-10.5 | Colourless | Blue |
| 28. | Tropeolin | 11.0-12.7 | Yellow | Orange Brown |
| 29. | Piorrier's blue | 11.0-13.0 | Blue | Violet pink |
| 30. | Indigo Carmine | 11.6-14.0 | Blue | Yellow |

## BY pH METER

This instrument is based on principle of Redox potential. By these instruments, accurate pH measurement of aqueous sample, paste and semisolids can be measured between range of 0-14. Different kinds of pH meters are available in the market, with meter digital readings, graphs etc. pH meters are also available for field investigation.

Commonly used laboratory pH meter is a double electrode meter, having a glass electrode and a reference electrode, separately. In some instruments, both the electrodes are combined in a glass cabinet, gives appearance as a single electrode instruments.

## OPERATING INSTRUCTIONS

1. There is a main measuring unit, housed in a cabinet with A.C. regulator, A to D convertor LSI and various segments display facility.

2. There are two main elements

   a) Glass electrode and

   b) Reference electrode

   Both the electrodes are built with blunt terminals and thin end chord

3. Both the electrodes are fixed in a holder and placed in a stand.

## PREPARING THE INSTRUMENTS FOR ANALYSIS

Before using the instrument (Fig. 2.1 to 2.3) of analytical work, some of the following preparation are required. They are as follow :-

1. Dip the pH sensitive tip of the glass electrodes in distilled water for atleast 4-6 hours in order to obtain stable and of course, reproducible, dependable readings.

2. Check the pottasium chloride solution's level of the reference electrode and if it is below the level, which is to be required, unscrew the cap of the electrode, existing at the top of the same and fill it with saturated pottasium chloride solution. This chemical should be of analytical grade.

   It is most important to notice that - for stable readings, it is essential to keep the hole open of the reference electrode open during measurement.

3. Adjust the electrode on the clip and hork the clip to stand.

4. Insert the terminal jack of glass electrode in the rear part of the main measuring unit (marked as GE) and push it and make sure the terminal is fit perfectly. Then tighten the screw.

5. Insert the terminal of reference electrode in the rear part of main measuring unit at its corresponding place (marked as RE).

6. Now, connect the power chord to the main A.C. If there pin socket is not available, then use two-pin plug, but the third wire 'GREEN' should be connected to any water pipe. Now, the instrument is ready for use.

## MEASUREMENT OF pH

To measure the pH of the given aqueous sample, following steps are sequentially required-

(a) ***Standardization of the instrument***

The instrument is standardised against known buffer solution at ambient temperature.

(b) ***pH Measurement***

As soon as the instrument is standardised for buffer solution, the same is ready for measuring the pH of unknown given sample.

Again, it is important to notice that take care in the setting the temperature compensate and standardized control.

Similarly, only use standard buffer solution(s) of accurate value(s).

Details of both the steps are as follows:-

**(a) Standardization steps**

Keep the switch norm/auto-off (exists at the back of the instrument) at auto-off position, when standardizing the instrument.

1. Turn the switch No. 2 clockwise till it clicks on.
2. Put the side switch No. 1 to 100% slop position.
3. Press the switch No. 4B (stand by position).
4. Rinse the electrode with distilled water and wipe them carefully with soft tissue paper.
5. Take the standard buffer solution (freshly prepared), say pH 7.0, in a beaker and record its temperature.
6. Dip both the electrodes in the buffer solution.
7. Set the temperature control knob-5-at the same temperature, recorded for buffer solution in proceeding step No. 5.
8. Press the function switch No. 4 at pH position.
9. Now, try to adjust the buffer value on the digital screw to read 7.0 by rotating the standardised control knob No. 3 either clockwise or anticlockwise, depending upon display.

Knob 3 (for Stardardise) is provided with a lock-nut-loosen, which can be loosen by hand and then try to operate the knob by inserting a small screw in the slit. As soon as, the unit is become standardised, tight on the nut so that the standardisation is not disturb.

Precaution should be taken to do not disturb the standardize control till at all the measurements are completed.

**(b) pH measurement of unknown solution**

To measure the pH of unknown aqueous solution, following steps are involved:-

1. After the standardization has been done, take out the electrodes from the buffer solution.
2. Wash the electrodes with a jet of distilled water and wipe them thoroughly and carefully with soft tissue paper.
3. Take the aqueous sample in a beaker, record the temperature of this unknown solution and set the temperature control to this corresponding values.
4. Press the switch No. 4 (pH) and read the pH of the unknown solution on digital display. Wait for minute. As soon as display shows stable reading, not the reading and give a gentle movement to the unknown solution and check the readings, again. As soon as, readings become stable, note it down in record-book.

## APPLICATION AREAS

It can be seen in the following application areas:-

(a) Preparation of standard buffer solution for chemical, biochemical studies.

(b) Standardization of an alkali or an acid solution.

(c) Determination of $P_{Ka}$ values of acids.

(d) Determination of concentration of strong acids & their mixture.

(e) Study of reactions involving liberation of hydrogen ions.

(f) Verification of Hederson's equation.

## APPARATUS AND CHEMICALS INVOLVED IN pH MEASUREMENT

Following apparatus and chemicals are involved in pH measurement of unknown solution:-

(i) ***Chemicals***

(a) Ortho phosphoric acid ($H_3PO_4$)

(b) Pottasium hydroxide (KOH)

(c) Oxalic acid $(COOH)_2$

(d) Distilled water

(e) Buffer solutions of pH 7, 9, 4, etc.

(ii) ***Apparatus***

(a) pH meter

(b) Beaker(s)

(c) Pipettes

(d) Stirring glass rod

(e) Wash bottle

(f) Appropriate civil and electrical arrangements.

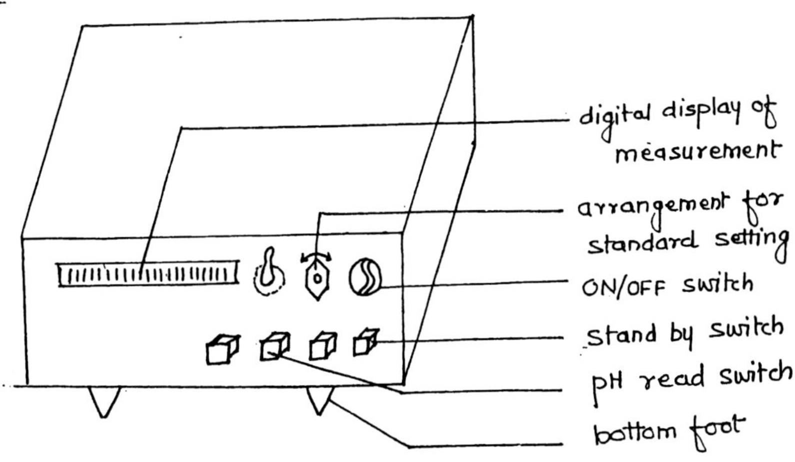


**Fig. 2.1 : Main unit front view**

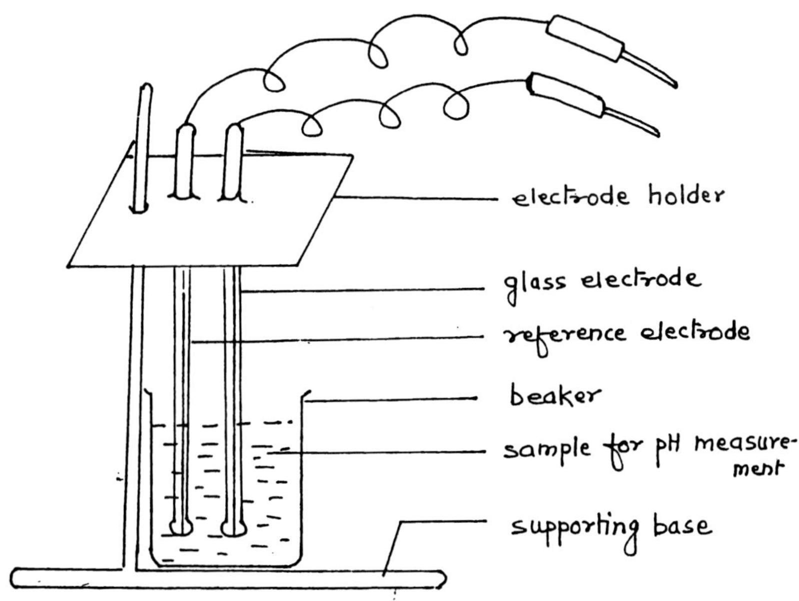


**Fig. 2.2 : Electrodes' arrangement for pH measurement**

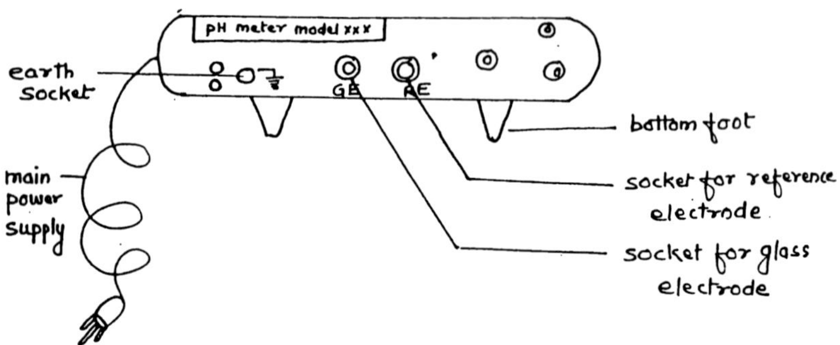


**Fig. 2.3 : Main unit real view**

# Chapter -3

# HOT PLATE CUM MAGNETIC STIRRER

## COMPONENT UNITS

### 1. Main cabinet

It contains main unit for magnetic stirrer & a heating device. Each of them has separate control on the main cabinet.

### 2. Hot plate

It is fixed at the top of the cabinet. It is a rectangular metallic plate, which is fixed with the help of porcelain spacers.

In between the plates and main cabinet, there is a radiation shield (Fig. 3.1).

## OPERATION

1. Keep the beaker with the solution to be stirred on the hot plate.
2. Leave the magnetic stirrer paddle (a small cylindrical iron piece supplied separately) gently into the solution & at the centre of the beaker.
3. Clamp the beaker with a stand, if required.
4. Connect the chord of the cabinet to main & switch it on. The red bulb of the right hand top was glow. If it does not glow, then proceed to stop and see that the bulb glows.
5. Switch on the heating knob (1) of the right side by moving it clockwise & fix it at a desired temperature

6. Put the switch No. 2 at ON position to start stirring.
7. Control the stirring rate by moving the switch No. 3 of the left hand side.
8. After the desired period, put off the heating knob and stirring switch and then the mains.

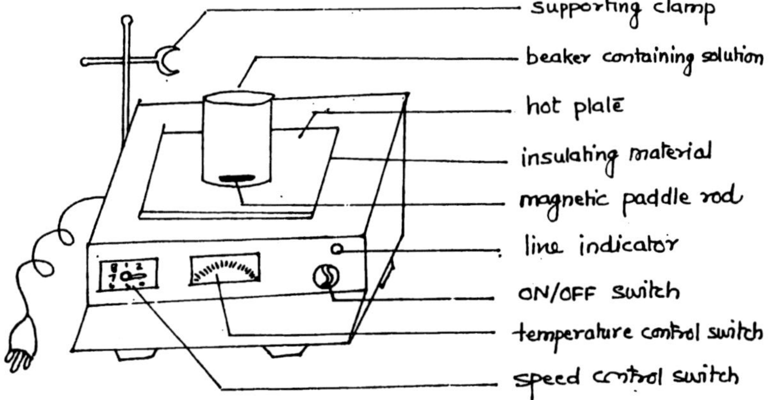


**Fig. 3.1 : Hot plate cum magnetic stirrer**

## Chapter -4

# DENSITOMETER

## INTRODUCTION

This instrument is used for scanning chromatogram, photograms on paper strips, glass slides etc.

The instrument uses to a directly calibrated sensitive micrometer. The light source, the slit and the optical filters are fixed in such a manner that the optical system remains in alignment.

The scanning frame can be used for moving the sample (filter paper strip) in precise 1 mm steps. The unit has a built-in stabilizer & operates on normal 220-240 volts A.C. mains.

Green and orange filters are also provided with the unit, which are used according to the selective nature of the chromatogram or electrophoresis, paper strips.

## AREA OF APPLICATION

It is mostly used for measuring the optical density of electrophoresis strip or any other semi-transparent films.

## PRINCIPLE

Paper strip electrophoresis is carried out in an electrophoresis apparatus. The strip, thus obtained are carefully dyed to render the separated bands clearly visible.

The ready strip is placed between the glass plates and the plates are mounted in the carrier. The carrier is inserted in the light path. Apart of light is observed by the strip and falls on a photoelectric cell.

The current obtained from the cell which is a measure of optical density is amplified & indicated on a meter (Fig. 4.1).

## OPERATING INSTRUCTIONS

1. Take the prepared electrophoresis strip and carefully place it on one of the glass plate & cover it with other plate.
2. Place the pair of glass plates in the carrier in such a way that reading edge is in the direction marked insert.
3. Insert the carriage (with the graduated scale facing the operator) into the slot on the left side sufficiently.
4. Rotate the knob No. 1 the clockwise direction.
5. Switch on the instrument (No. 2, Zero control).
6. Insert the blank plate in the filter holder.
7. Adjust the zero control No. 2 & set meter to zero on the left of the meter scale.
8. Wait few minutes for instrument to stabilise, check 'O' again.
9. Replace the blank with an appropriate colour filter.
10. Advance the carriage so that you see zero millimetre on the scale is in line with reference line.
11. Set the meter to 100 transmission (extreme right of meter) by means of set 100 control No. 3.
12. Now, the instrument is ready for scanning.
13. Advance the carriage (with the strip in position) millimetre by millimetre & note down the meter reading either on percentage transmission scale or O.D. Scale.
14. Plot the readings on a graph sheet.

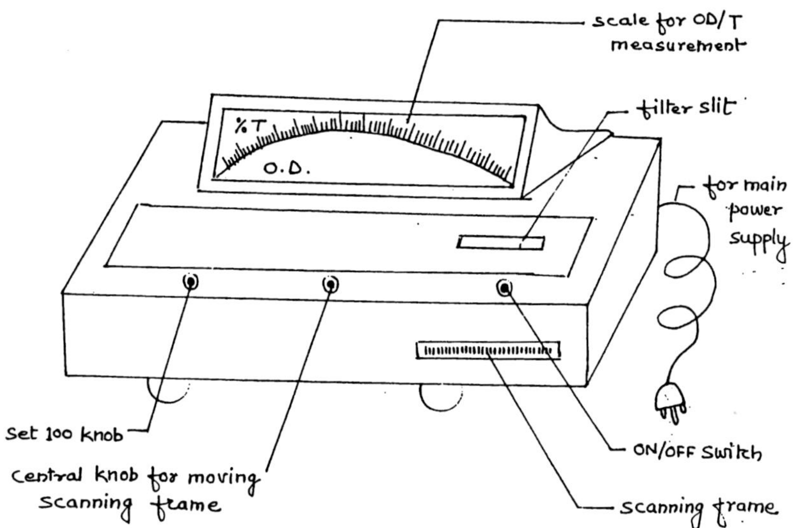


**Fig. 4.1 : Densitometer**

## Chapter -5

# CONDUCTIVITY METER AND MEASUREMENT CONDUCTIVITY

## INTRODUCTION

Conductometry involves the measurement of the conduction of the solution of electrolytes (ionic substances). Being an additive property, this can also be used (similar to pH metry) for determination of the concentration of acid or base in solution, dissociation constants of acids or tracing the reactions between the ionic substances.

## PRINCIPLE

Ohm's law stated that – "*the Current (in ampere) flowing in a conductor is directly proportional to the applied electromotive force E (in volts) and universally proportional to the resistance R (in ohms) of the conductor.*"

Mathematically, it can be expressed as :

$i \, \alpha \, E/R$

The reciprocal of the resistance is termed the *conductivity of the conductor.*

The resistance of a homogeneous material of uniform cross section with an area of 1 Sq. c.m., the length 1 c.m. is given by-

$$R = \frac{P.1}{a}$$

Where

P = characteristic property of the material termed as ***'specific resistance'***.

The reciprocal of the specific resistance is named as the specific conductance 'K'.

This is the conductance of a cube of material having 1 c.m. length and 1 c.m. in cross section. The specific conductance of an electrolyte an any temperature, depends only on the ions present and, therefore, varies with their concentration.

If all the solution be placed between two electrodes, 1 c.m. apart and large enough to contain the whole of the solution, the conductance will increase as the dilution increases.

If 1 gm equivalent of the solute is present, the conductance of such a such a solution is termed ***'equivalent conductivity'***.

Thus,

$$\lambda = L.V$$

Where V = Volume in millilitre

Containing 1 gm equivalent of the solute.

The addition of an electrolyte to a solution of another electrolyte under conditions producing no appreciable change in the volume, will affect the conductance of the solution, according to whether or not ionic reactions occur.

If no ionic interaction takes place, such as in the addition of one salt to another (e.g. KCl to $NaNO_3$), the conductance will rise.

If ionic interaction occurs, the conductance usually changes. Thus, in the addition of a base to a strong acid, the conductance decreases owing to the replacement of hydrogen ion (high mobility) by another cation (lower mobility).

This is the principle underlying conductometric titration, i.e. the substitution of ions of one mobility by ions of another mobility.

## DESCRIPTION OF INSTRUMENT

The instrument used generally consists in a direct reading conductivity bride using a conductivity cell. A conductivity cell consists in two platinum electrodes 1 square c.m. in area, welded to thick platinum wires fused into pyrex glass tube.

These make contact with mercury

Bright platinum electrodes should be platinised in order to minimise polarisation effects.

When absolute value of the conductance is required, the So called cell constant is determined. If the resistance of the liquid measured in the cell is R and the specific conductance is K, the cell constant is defined by the following relationship-

$$K = \frac{\text{Cell constant}}{R}$$

or Cell constant = K.R.

It is easy to calculate the cell constant from the measurement of the conductance (or resistance) of a solution of known ***"specific conductance"***.

Potassium chloride solution are generally employed since the specific conductance of aqueous solution of this salt have been determined with a high degree of accuracy at several temperature. However, for conductometric titration, it is not essential to know the absolute conductance.

## EXPERIMENTAL PROCEDURE

Determination of stoichiometry of HgI complex.

## APPARATUS

Conductivity bridge, two standard sized flasks, beaker, pipette, burette, glass rod etc.

## CHEMICALS & SOLUTIONS

Mercury chloride, potassium iodide, distilled water. Prepare 0.01 M solution of mercury chloride and potassium iodide.

## PROCEDURE

Pipette 5ml. of 0.01 M mercury chloride solution in 100 ml. beaker and to this, add 15 ml. of distilled water.

Measure the conductivity of the solution.

Now, titrate this solution with 0.01M KI solution and measure the conductivity after the addition of 1 ml. of KI solution.

Plot the titration curve (conductivity versus millilitre of potassium iodide added & explain the curve.

The reaction between $HgCl_2$ and KI takes places follows:-

$HgCl_2$ + 2 KI ———> $HgI_2$ (scarlet red ppt)

$HgI_2$ + 2 KI ———> $K_2HgI_4$ (colourless solution)

So, during the titration, first the scarlet red coloured precipitate of mercuric iodide will be formed, which will then start dissolving due to the formation of colourless $K_2HgI_4$ complex.

Hence, in the curve, there will be two breaks, one at 10 ml. of KI and the other at 20 ml. of KI, showing the formation of $HgI_2$ and $K_2HgI_4$.

## AREAS OF APPLICATION

1. Concentrations of acids or base.
2. Dissociation constant of acids.
3. Tracing the reaction between the ionic substances.
4. Determination of electrolyte concentration in a sample of water.

## Chapter -6

# CENTRIFUGATION

## INTRODUCTION

It is a common knowledge that when insoluble particles are suspended in a aqueous medium, the larger particles settle down at faster rate i.e. the sedimentation of larger particles is faster.

The process of sedimentation is caused due to the gravitational force of the earth.

Based on the similar principle of gravitational force, the device of artificial production of gravitational force much higher than the earth's gravitational force is introduced with the help of centrifuge machines. There has been a gradual evolution in the centrifuge machine, with respect to their revolution capacity, ranging from 1000 g in clinical centrifuges to 1,50,000 g in ultracentrifuge (Fig. 6.1 and 6.2).

## CLASSIFICATION

The centrifuges may be grouped in following categories :

(1) ***Low RPM*** **(Revolution Per Minute)**

Generally clinical centrifuges are of this category. It has maximum revolution upto 8000 RPM. Commonly these are used for general clinical, chemical, biochemical, microbiological studies of environmental samples for centrifugation of preliminary nature.

(2) ***Medium RPM***

Basically this type centrifuges are refrigerated in nature. It has maximum 20000 RPM and commonly used for sedimenting bacteria and cell organelles viz. mictochondria etc.

(3) ***High RPM***

These are known as ultracentrifuge. It has maximum 150000 RPM. It is commonly used in studies related to wastewater microbiology, environmental science and engg., public health engg., biochemical engg. and genetic engg. etc.

**PRINCIPLE**

The rate of centrifugation (or sedimentation) of the suspended particles in a aqueous medium depends on the following factors:-

1. The magnitude of centrifugal force.
2. Duration of centrifugation.
3. Size and density fo the particles.
4. The medium of suspension.

These factors may be explained in following manner to evaluate the extent of sedimentation of the particles in a aqueous environment:

**(a)** ***The magnitude of the force***

The magnitude of the centrifugal force, again depends on the following variables:

(i) the length of the tube i.e. the distance of the sedimenting particles to the centre of the axis. and,

(ii) the number of revolutions per minute (RPM).

The magnitude of the centrifugal force is expressed in terms of the earth's gravitational force 'g' (which is 980 C.M./Sec$^2$ in C.G.S. sysem)

Mathematically, it can be expressed by following formula-

$$F = \frac{S^2 r}{89500}$$

Where,

F = Centrifugal force in 'g' units.

S = revolutions per minute and,

t = radius (distance between axis to the end of the tube).

(b) ***Duration of centrifugation***

It is a also important factor. In general, longer the time, more the sedimentation (quantitatively) occurs.

(c) ***Size and density of the particles***

The rate of sedimentation depends on the size and density of the suspended particles. The denser and larger particles have fast rate of the sedimentation.

(d) ***Density of the medium***

The density of the suspending medium, in which particles are suspended, play an important role in the sedimentation of the particles. The velocity of the sedimentation is inversely related to the density of the medium.

Mathematically,

$$\text{Sedimentation} \propto \frac{1}{\text{density of the medium}}$$

It says, if the suspending medium has less density, more sedimentation takes place.

## ADVANTAGE OF DENSITY GRADIENT

The phenomenon of density gradation helps in the separation of the particles of different size and density. This is due to the phenomenon that the centrifuge tube, the density of the medium, in which particles of different size and density are suspended, increases from top to bottom either stepwise or linearly.

When this suspension is subjected to centrifugation, the particles are starts sedimentation. When they reach a point, where the density of them equals the density of the medium, they remain in that position, even if the centrifugation is continued. In this way, different layers of various substances can be noticed along the tube.

## PROCEDURE

1. The mixture of powdered substance is suspended in water of known quantity in the tube. If the fractional substances of the tissue are to be separated, it has to be homogenised either in an electric grinder or with pastel and mortar in the following way-

Take the tissue in the grinding vessel containing 0.4M sucrose solution at the ratio of 5 ml. per gm per fresh tissue weight .

Filter the homogenate through double or four layered muslin cloth and store the filtrate in the cool place.

2. Measure the radius (r) in centimeter from the centre of the axis to the circle in which the tube is to rotate.

3. Adjust the RPM at which the centrifuge is to be driven and calçulate the 'g' force at this RPM.

4. Insert tube (with the sample in it) in the rotor and insert another tube containing the same amount medium, say water, in the opposite holder of the rotor (for balancing the tube during operation).

5. Close the lid and switch on the centrifuge. If there is a clock arrangement in the instrument, adjust the knob at the desired timing.

6. Allow the centrifuge to run for the desired period and switch off the current.

7. Allow the centrifuge to stop by itself. Don't try to stop by hand.

8. Open the lid and remove the sample tube. Decant the supernatant solution carefully into the clear beaker or tube. The solid pallet left at the bottom of the tube represents the solid particles or particulate matter that would sediment in the force field developed at this RPM.

   The supernatant solution taken out in the beaker may contain particles which would be lighter and less dense to the sedimented at this RPM. They would have require more RPM for sedimentation.

9. Pour 2 ml. distilled water in the tube and dissolve the pallet through shaking. Take a drop of it and examine under microscope.

**EXPERIMENTATION**

1. Pour the supernatant isolation (obtained in the step 8) into a clean tube and repeat the steps 3 and 5. Adjust the RPM 500 and keep the same time of centrifugation. Follow the steps 6-10.

Correlate the size of particles sedimented in the two different magnitudes of the force fields developed at two different RPM speeds.

2. Find out the results of centrifugation by using different medium, say glycerol. Take equal volume of glycerol in two centrifuge tubes and add 1 ml. of supernatant solution obtained in step 9 to one of these tubes and to the other. Add 1 ml. of water for balancing the sample in the rotor.

   Run the centrifuge at the same RPM for same period and compare the size of the particles in the two samples.

3. Compare the results by increasing the duration of the centrifugation with the same medium.

4. Compare the results by increasing RPM and keeping the same time period.

5. Compare the results by increasing the RPM as well as duration.

## DENSITY GRADIENT CENTRIFUGATION

There are two methods of preparing the density gradients:-

a) Discontinuous gradient and,

b) Continuous gradient.

It is explained in following manner :

a) ***Discontinuous gradient*** :

Prepare different grades of M sucrose solution say 10%, 30%, 50%, 70%, 90% Take 10 ml. centrifuge tube and pipette out 2 ml. of 90% solution into it.

See that all the drops have reached the bottom. Then pipette out 2 ml. of 70% M sucrose solution, carefully, through the wall of the tube, so that the bottom layer of 90% sucrose is not disturbed. Similarly, pipette out 50% then 30% then 30% sucrose solution, each two ml. in same way to avoid any stirring of the previous layers. Different layers can be seen due to the differences in their refractive indices, when viewed against light.

b) ***Continuous gradient :***

1. Take 5 ml. of the homogenate of the tissue already prepared (in a pipette) and layer it carefully on the top of continuous gradient.

2. Insert this tube in a swing out rotor so that the sample remain in horizontal position during rotation.

3. Balance it by inserting another tube with equal amount of water in opposite side of the rotor.

4. Centrifuge the tubes at 500 RPM for 15 minutes and observe the results are as discussed earlier.

Repeat the experiment with continuous gradient at different RPM and time duration.

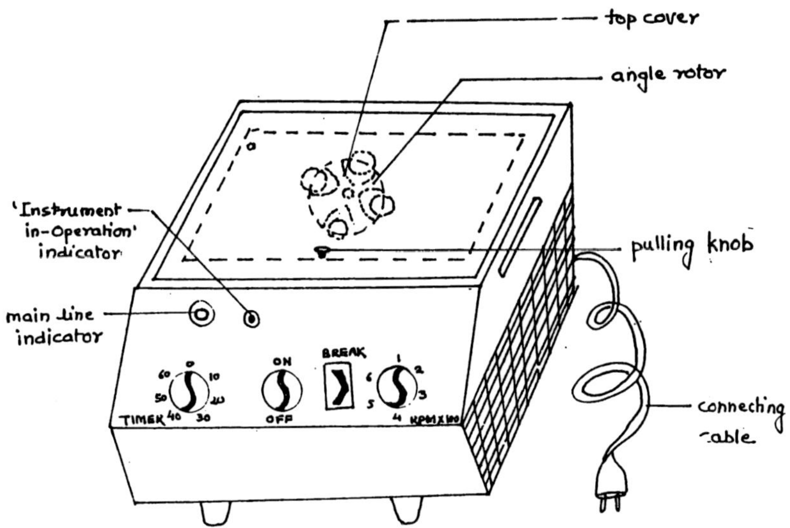


Fig. 6.1 : Centrifuge main unit

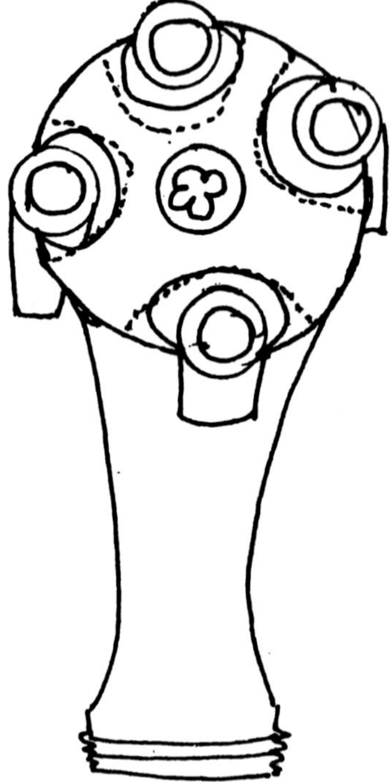

**Fig. 6.2 : Centrifuge angle rotor**

## Chapter -7

# REFRIGERATED CENTRIFUGE

## INTRODUCTION

It operates at speed upto 16000 RPM and a maximum gravitational field of 23000 g with high separation capacity. The lowest temperature in the centrifuge cup can be attained to -6°C and thus all temperature sensitive objects can be handled without any loss. (Fig. 7.1)

## AREA OF APPLICATION

This instrument can be used in following disciplines :

a. ***Bacteriology***

Bacteriological serology for sedimentation and enrichment of bacteria.

b. ***Medicine and veterinary medical virology***

for sedimentation and enrichment of viruses (bacteriophages).

c. ***Physiological chemistry***

cell fermentation.

d. ***Plant virology***

Sedimentation and enrichment of the largest rod shaped viruses.

## COMPONENT UNITS OF CENTRIFUGE

1. ***Cooling unit***

   This unit is lodged inside the main cabinet and consists of evaporator and an air tight cooling unit. The cooling unit is arranged in the lower part of the centrifuge housing and is swing mounted.

2. ***Centrifuge drive***

   The driving motor is housed in the centrifuge housing in two planes of oscillation. The conical rotor, a speedometer is accommodated in the lower shaft end of the driving motor ensuring speed control in connection with a speed indicator. An ampere-meter is built in for controlling the current. These are three press keys to control the motor operation-

   1. One for drive (I)
   2. The other for clock or timing (t) and the
   3. for brake (O)

   Generally, corresponding symbols of these switches given on the panel of the centrifuge.

3. ***Rotors and accessories***

   There are three black coloured angular rotors with different sizes of holes to accommodate the tubes. The rotors are equipped in the works with plastic cups. Plastic cups are suitable to withstand the strain of high speed.

## OPERATIONAL STEPS

### Installation of rotor

1. Press the steel knob (No. 9) of the cover at the top and raise the cover from the right to left aide in order to open rotor chamber of the instrument.
2. Put the rotor on the catch in such a way that the 2 pins of the catch are insert in the holes in the holes of the rotor.
3. The plastic cups filled with the material to be centrifuged, should be symmetrically arranged in the rotor.

## Operation

1. Press cooling switch (No. 3) of the instrument to start the cooling device.
2. The desired temperature is set with the help of metallic key (provided separately) by rotating the metallic needles of refrigeration meter (No. 6). The black needle of this meter will move towards the desired temperature in course of time.
3. Depress the ON/OFF (No. 1) switch to drive the motor and run the centrifuge.
4. Before it, adjust the clock for which, centrifuge is required to run.
5. Turn the black large dial knob (on the front panel) of the regulating transformer to the zero position and thus start meter.
6. Note the ampere meter reading on the meter. It should not be exceed 4 ampere in starting.
7. Adjust revolutions on RPM dial (No. 8) by the same large knob.
8. Press the bottom (No. 2) (t) for timing and simultaneously press the button (No. 4) (O) for putting a break after the desired time. Thence the motor will stop automatically by the brake after desired period.

   The ON/OFF switch No. 1 can be put to off at the end.

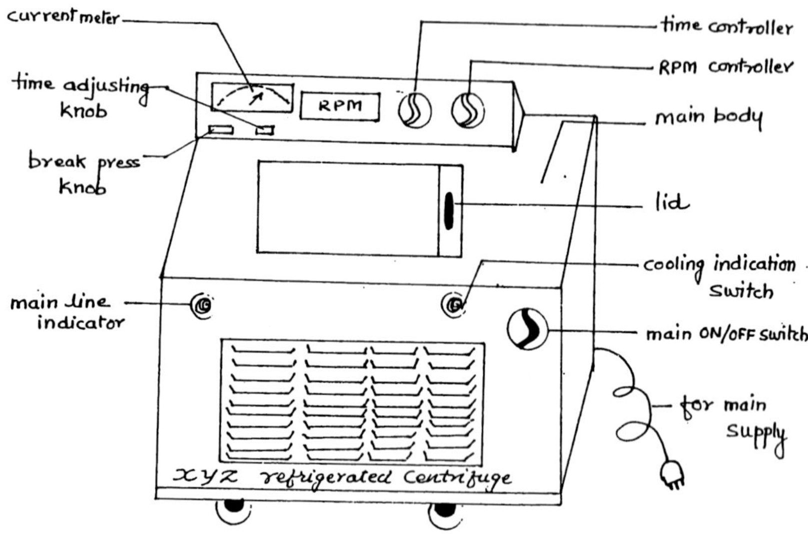


**Fig. 7.1 : Refrigerated centrifuge**

Chapter -8

# PHOTOMETRIC ANALYSIS : FUNDAMENTALS

## COLORIMETRY & SPECTROPHOTOMETRY

The atom of an element consists of a compact nucleus comprising positively charges protons & neutral neutrons.

The atomic number of an element is the same as the total number of protons & is characteristic of that element. Electrons, negatively charged particles of equal though opposite charge to that of protons in a neutral atom, surrounds, the compact nucleus in a more or less diffuse but orderly manner.

They may be thought of as located in orbits with which discrete energies are associated. Each orbit is spoken of as an energy level in the atom.

According to the quantum theory, radiant energy is emitted as quanta. The energy of each quantum is located to the frequency of the radiant energy "e" according to the following equation:-

$$e = h.v.$$

Where

h = Plank's constant ($6.62 \times 10^{-27}$ arg/sec.)

v = frequency of radiation in vibration/sec.

In both the visible and ultraviolent regions of the spectrum, which extend for ordinary analytical purposes from a wavelength of about 200 nm to about 400 nm for the ultraviolet region & from 400 nm to about 700 nm. for visible region.

In both regions the absorption of radiation is caused by the transfer of the energy that radiation to electrons in the outer orbit of the molecule which then suffers displacement.

In the visible region, the energy involved in this displacement is less than that in the ultravionent region but qualitatively the displacements are of the same type.

The absorption process is thus different from that which occurs in the infrared region, where the absorptions are governed by oscillations & displacements of atoms within the molecule.

The basic equation for absorption can be given as :

$$I = I_{oc}^{-abc}$$

## COLORIMETRY

In calorimetric techniques reactions are carried out which usually convert colourless substances to coloured, the colour serving as a means of identification & its intensity as a quantitative measure of the amount of the substance undergoing reaction.

## SPECTROPHOTOMETRY

It may be considered as the study of the relative measurement of radiant energy as a function of wavelength. The term relative should be stressed since in spectrophotmetry the radiant energy transmitted or reflected by a system is nearly always measured, not in terms of absolute units, but as related to some standard such as the solvent in which the substance being analysed is dissolved. It may be used to study fundamental properties of matters, such as molecular or atomic structure, it may be used to study and standardise colour as such & it may be used to identify the substances, either in their pure state or in mixture & to determine their concentrations.

White light from the Tungston lamp is focused by lens A on the entrance slit, lens B collect the light from the entrance slit and reflected & dispersed by the diffraction grating.

To obtain various wavelengths, the grating is rotated by means of an arm which is moved. when the camera is rotated, the wavelength scale is fastened to the sale shaft as the camera.

The monochromatic light which passes through the exit slit goes on through the sample to be measured and falls upon the photo tube, whenever the sample is removed from the instruments, the holder falls into the light beam so that the amplifier control can be adjusted with no further manipulation.

A light control is provided for setting the meter to full scale deflection with a suitable blank in the sample compartment. Covettess or special small test tubes are used as containers for the samples.

The apparatus although designed primarily as a calorimeter also serves as an inexpensive spectrophotometer.

## ULTRAVIOLET AND VISIBLE SPECTRO-PHOTOMETER

This is a precision instrument. Two interchangeable light sources are used, a tungston filament lamp and a hydrogen discharge lamp, the former for measurement down to 350 nm & the latter for measurement in the ultraviolent region below 350 nm. It employs a fused silica of the Littrow type with a concave mirror of 50 cm focal length for collimation.

The slit mechanism is continuously adjustable from 0.01 to 2.0 mm. Slit widths are ready directly from a calibrated dial and are reproducible to within 0.1 percent.

The wavelength range is from 210 to 1000 nm and wavelength scale readings are accurate to better than 0.5 nm. The blank spread of the monochromator can, if necessary, be adjusted to less than 1 nm with the appropriate setting of the slit opening over most of the spectral range of the instrument.

Two phototubes are employed, a red sensitive phototubes for use above 600 nm & a blue sensitive phototube for use above 600 nm and a blue sensitive phototube for use in the range 320-625nm (tungsten lamp) and 210-360 nm (hydrogen lamp).

The phototube current is measured by a null method utilising a slidewire potentiometer and an electronic amplifier. The potentiometer is calibrated in percent transmission from 0 to 100 percent and in optical density from 0 to 2. Controls are provided inter alia for adjusting the dark current to zero & the percentage transmission to 100.

Four standard rectangular absorption cells of 10 mm light path are supplied in a four-place cell holder, an interchangeable cell compartment is available to accommodate either pyrex or silica cells with path length from 2 to 100 nm and with sample volume from 0.3 to 285 ml.

## DEVELOPING COLORIMETRIC METHOD

Following important factors are required to study the development of a proper colorimetric method :

### Wavelength Selection.

In case of a colorimeter by way of introducing a choiced filter after source and before the susbstance.

### Interferences

By the substances that may be commonly associated with that under study and to remove them by way of precipitation or some other suitable methods.

### Effect of pH

If pH changes, the colour may change as the compelled formation itself will change and so a proper pH range should be found out.

### Absorbance concentration curve & its linearity

That where the curve maintains the linearity & find a portion where the deviation occurs.

### Reagent & their effects

To find out the minimum amount of reagent required for the development of colour.

### Factors that introduce fading

Time, heat, light etc. are some of the factors that introduce fading of reaction mixture.

### The accuracy of the method

This can be found once by following, we fixed all reagents, time, pH & other factors, the different concentration taken, the curve be, plotted and the known amounts be added to the standards & the recoveries be done.

Detailed explainations of each of the above mentioned factors are as follows :

**1. Wavelength selection**

It is required to first find out the absorption spectrum of the reagent blank (in most of the case, may be distilled water), then, after adding the reagents to a chosen concentration of standard depend colour & obtain the absorption spectrum.

In aqueous system, distilled water is taken as zero and when using the organic solvents, the solvent are taken as blank and adjusted to zero absorbance (fig. 8.1). $KMnO_4$ spectrum are presented in acidic medium.

***Taking the readings***

From the above spectrum, we can see that upto about 460 nm, there is hardly any absorption, but beyond this the absorption starts so that upto 460 nm, we can have about three to four readings, but above this, the readings, should be closely spaced by 5 nm difference or so to get a complete idea of the absorbance curve & above 580 again, we can have about three to four readings, that will give complete picture o the absorption curve.

Sometimes reagent blank is adjusted to zero. This is to check against the purity of the solvent and many other errors associated with the cleaning of cells.

**2. Interferences**

The interferences can be chemical or optical. If chemical, they can be removed by way of precipitation or by forming suitable complex that can be separated out.

In any case the absorption due to interference should not increase more than 10 percent & should be a constant throughout the procedure in all samples.

The interference can be corrected, if the substances are so closely having the absorption maxima such as in the following case of neodymium and Praseodymium dissolved in dilute HCl giving in the curves.

There are two substances that are interfering mutually. If now both are present in the solution & one of them is to be evaluated. Let us find out absorption at $A_1$ = 575 nm & $A_2$ = 590 nm.

$A_1 = a_{Nd\,(575)}c\ Nd^b + ma_{Pr\,(575)}{}^{C}\ Pr^b$

$A_2 = a_{Pr\,(590)}{}^{c}\ Pr^b + a_{Nd\,(590)cNd\,(596)}{}^{b}$

Where

b = path length known

a = absorption measured by taking pure substance

c = concentration taken

## EFFECT OF pH

In all these methods, the pH of the solution is critical. That can be better illustrated by taking the pH papers & how the colour changes. Hence the pH change does not effect on the absorption pattern but the intensity of colour changes with pH variation & so absorption values may change.

Thus in developing a calorimetric method, one has to select a rather broad range of pH & that small errors in pH do not effect the absorption of substance appreciable.

## AREAS OF APPLICATION

Absorption technique is very useful than the emission technique. Because emission technique is cumbersome & takes long time & the instrumental parts is very complicated. On the other hand, in absorption technique, the instrumentation is very simple and considerably time saving.

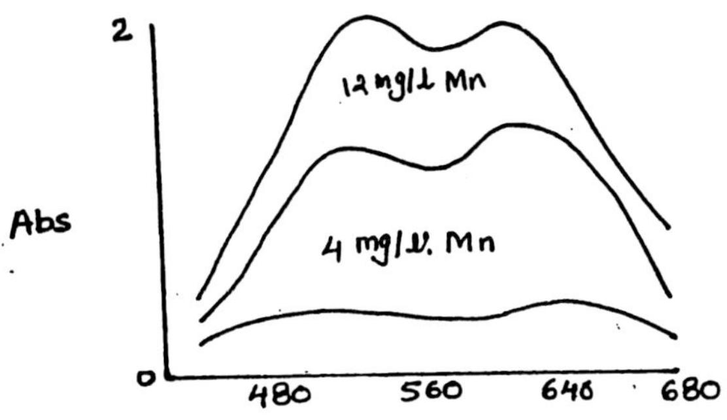


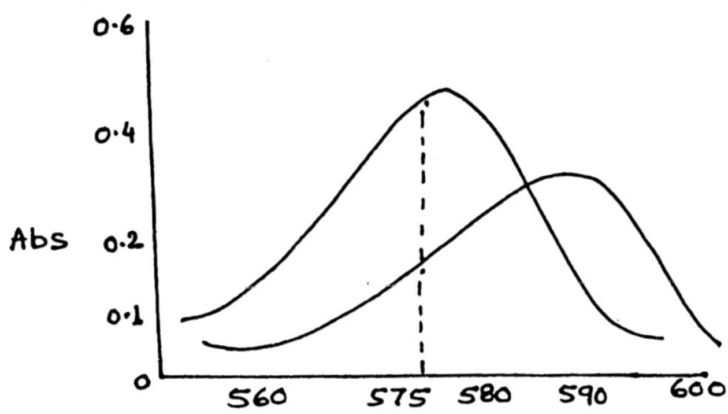


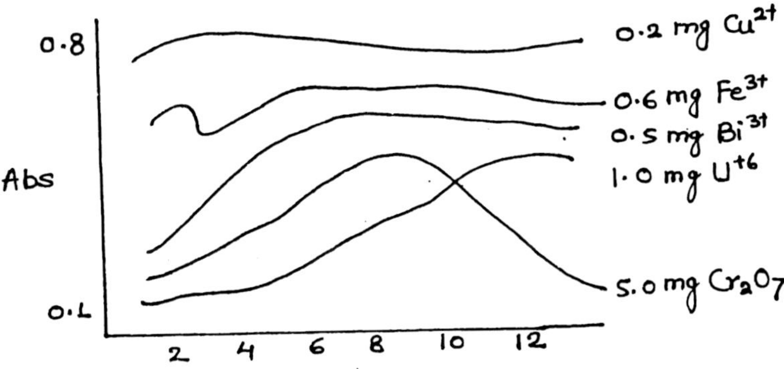


**Fig. 8.1 : Wavelength in photometric analysis**

# Chapter -9

# COLORIMETRY

## INTRODUCTION

The colorimeters are widely used in chemical, biological and engineering researches, especially in determining the concentration of substance(s), tissue extracts preparation etc. This is done by determining the optical density of the solution. This instrument can be used in variety of colorimetric analysis in the field of engineering, medicines, environmental & biotechnological studies, foodstuffs, agriculture, plastic, paints, textile, paper, soap and other industries like leather, brewing, gas etc.

*Limitations* : Colorometers are used for routine work of estimations & reaction kinetics, because their range is limited to visible light. For highly accurate work, spectrophotometers are used.

Spectrophotometers are have some advantages over colorimeters, because in spectrophotometers, prisms are used instead of filter. It also has wavelengths accuracy of ± 0.1 μ can be achieved due to grating or prism arrangement. In addition, the spectrophotometer can also be used in the ultraviolet region (upto 240 nm wavelength).

## PRINCIPLE

All compounds absorb light of certain wavelength, depending on the chemical composition are characteristics of the said compound. This absorption is maximum at a particular wavelength and is known as ***"absorption maxima"***.

The absorption of light in a aqueous solution is governed by two fundamental laws. These are as follows :

(i) ***Lambert's Law***

*It states that* the amount of light transmitted through a aqueous solution, depends on the distance of light parh and,

(ii) ***Beer's law***

*It says that* The amount of light absorbed per unit thickness of the aqueous media is proportional to the amount of light absorbing particles, existing therein (i.e. conc. of the substances, present in the solution).

A schematic diagram of colorimeter given in Fig. 9.1 and 9.2.

## FACTORS INVOLVED IN COLORIMETRY

The various factors involved in the colorimetry may be represented as follows:-

$I_0$ = intensity of incident light

I = intensity of transmitted light

T = Transmittance (in %) $I_0/I$

L = distance through which light passes

C = concentration of the solution

Thus -

$$\log I_0/I = \log \frac{L}{H}$$

$$= K.L.C. = \text{optical density}$$

Where,

K = constant

and is called ***extension co-efficient***. K remains constant for all compounds under the same conditions of following-

1. Solvent
2. Temperature
3. Wavelength of light

## OPERATION

A few voltage temperature (6-6.5volts) energised by a stabilized supply from a light source (a). The light posses through a selected filter (b) and a slit (c) and then passes through the test tube, containing sample solution (d) and ultimately, falls on the sensitive photocell (e).

The current generated by this photocell (e), is amplified by an amplifier (f), which drives the galvanometer (g).

The galvanometer is calibrated both in term of :

(a) optical density and

(b) percentage transmittance

## SPECIFICATION OF FILTERS

The filters are given specific numbers and each number represents a particular wavelength. The filters are mounted on a point, so it can be able to rotate and any filter can be brought in the path of light by rotating the filter on the said point.

Generally, there are 8 filters are mounted in a frame with specifications, given in Table 9.1.

**Table 9.1 : Specification of various filters, present in a single frame**

| *Filter No.* | *Colour* | *Wavelength range (°A)* | *Peak transmittance (°A)* |
|---|---|---|---|
| 621 | violet | 6800-5100 | 4200 |
| 622 | blue | 4000-5300 | 4400 |
| 623 | blue-green | 4600-5400 | 4900 |
| 624 | green | 4900-5600 | 5200 |
| 625 | yellow | 5400-6100 | 5700 |
| 607 | orange | 5700-7000 | 6000 |
| 608 | red | 6300-above | 7000-above |

## PRINCIPLE OF COLOUR ABSORBANCE/ TRANSMITTANCE

A solution of a particular colour with a specific wavelength range transmits the rays of only that particular colour and absorbs all the other wavelengths.

For example, a solution of copper sulphate is blue because it transmits blue colour only and absorbs all other colours. It is explained more clearly in Table 9.2.

**Table 9.2 : Wavelength absorption & transmission by various coloured solution (s)**

| *Colour absorbed* | | *Colour transmitted* | |
|---|---|---|---|
| *Colour* | *Wavelength* | *Colour* | *Wavelength* |
| Violet | 400-435 μ | Yellowish green | 560-580 μ |
| Blue | 435-480 μ | yellow | 580-595 μ |
| Greenish blue | 480-490 μ | Orange | 695-610 μ |
| Bluish green | 490-500 μ | Red | 610-750 μ |
| Yellow | 580-595 μ | Blue | 435-480 μ |
| red | 610-750 μ | Bluish-green | 490-500 μ |

## OPERATIONAL PROCEDURE

To operate the instrument for analysis, following steps are involved :

1. Set the SET 100 (say knob 'B' control to maximum by rotating knob 'B' anticlockwise to the extreme left. Keep the LID closed (C).

2. Connect the instrument to 230 V, 50 HZ line through the chord 'f' in the rear of the instrument.

3. Switch on the instrument by rotating the knob B slightly clockwise and leave it to warm up.

   The instrument will stabilize after a few minutes.

4. Adjust the needle to "O.O" optical density (O.D.) with the help of knob A (Set 'O').

5. Raise the lid 'C' and insert the black solid tube (provided separately) in the sample holder aperture (D) and close the lid again.

6. See that the O.D. is otherwise read just that the needle to with the knob 'A' (set zero).

7. Select the filter by rotating the frame 'E' Selection of proper filter is the key of success.

8. Lift the lid and replace the black tube with a tube containing blank solution (usually distilled water). close the lid again. Care should be taken that the mark (white line) on the tube coincides with the mark (1) of the tube holder aperture.

9. Adjust the SET 100 control by rotating knob 'B' So that the meter read O on the optical density (or say 100 on the percentage transmittance) scale. This will indicate hundred percent transmission.

10. Lift the lid. Remove the blank solution and insert there black tube and observe is there any change. If it is so, then readjust 'SET O' by rotating knob A to optical density. Recheck optical density to 'O' by replacing blank solution in sample holder aperture. If no change is found, it means, the instrument is ready for use.

11. For testing the optical density and or percentage transmittance of a solution, the solution is put in the sample tube and the same is transferred to sample holder aperture in the place of blank solution. As soon as the pointer of meter, stops deflections, reading is noted and recorded.

## EXPERIMENTS TO BE PERFORMED

### (i) Complimentary colours

Make solution of different coloured substances like :

(a) methylene blue

(b) Pottasium dichromate ($K_2Cr_2O_7$)

(c) haemolysed blood

(d) tissue extract of biological origin

Measure the optical density (absorbency) of each using different filters. Record and interpret the observations.

### (ii) Beer's laws verifications

Take a pure compound like methylene blue & make a 10 mg/ml. sol. accurately Record and interpret the observations.

## AREAS OF APPLICATION

1. Estimation of a substance in solution.
2. Study of reaction kinetics.
3. Determination of stoichiometry of substances, formed in solution.
4. Determination of solubility of substance, formed in solution.

## ACCESSORIES USED IN COLORIMETRY

Following chemicals and apparatus are required for colorimetric studies

(i) ***Chemicals***

Distilled water and other chemicals, which are required for given experiment.

(ii) ***Apparatus***

In addition to calorimeter, micropipettes, test tubes, glass rods, wash bottle, volumetric flasks etc. also required as glass wares.

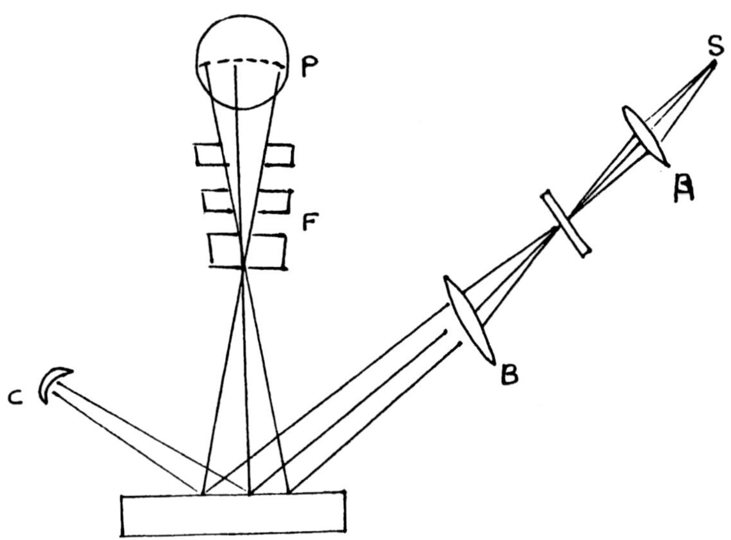


**Fig. 9.1 : Schematic of Spectronic - 20**

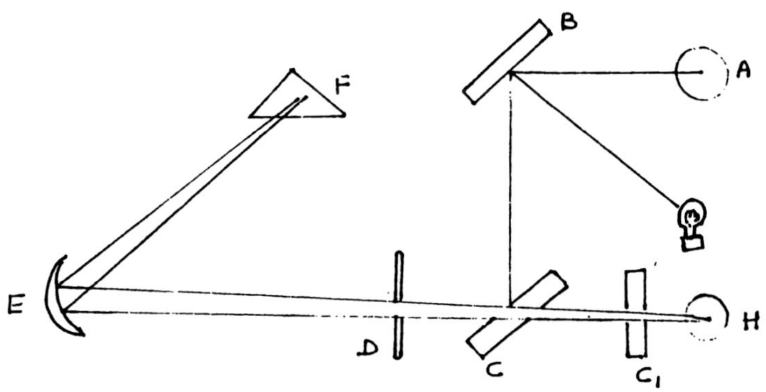


**Fig. 9.1 : Backman D.U.**

## Chapter -10

# SPECTROPHOTOMETRY

## GENERAL INTRODUCTION AND PRINCIPLE OF OPERATION

### Absorption of radiant energy

When a beam of monochromatic (single wave length) radiant energy falls on the homogenous layer of transparent substance, some energy is absorbed & the remainder is transmitted.

Actually, a small amount of energy is also reflected. Generally, the instrument used for making measurement is designed so that the factor councils out. If the incident radiant energy includes wavelengths in the visible region of the spectrum and the medium, through which, is passes can selectively absorb certain wavelengths, the visual colours observed, will correspond to the wavelengths of the energy transmitted.

In a simple way, a solution is observed as being blue because it absorbs the rest of the waves and transmits the blue waves. Similarly, a purple solution transmits only purple (blue and red portion of the spectrum) and absorbs strongly yellow-green in between.

It is within the experience of students in qualitative analysis that the intensity of colours of a given solution is observed to be differing when it viewed through different thickness & that when view through identical thickness, solution of different concentration of the same substance differ in intensity and observed colour.

These generalisations have a quantitative basis in the law of spectrophotometry.

## THE SPECTROPHOTOMETRY LAW

### (i) Bouger's Law:

It is also known as ***Lambert's law***, consists of two parts

(a) *The monochromatic radiant powers transmitted by a given homogeneous isotopic medium, is proportional to the incident radiant powers or the ratio of the transmitted radiant power P, to the incident radiant power $P_o$ is a constant*

Mathematically :

$$T = \frac{P}{Po}$$

Where T is a constant and is in fact the transmittance.

(b) *The transmitted radiation power decreases in geometric progression as the length of the optical path increases in arithmetic progression. It can be stated in different way - layer of equal thickness absorb equal fractions of the radiant power incident upon them. These relation may be made clear to those who, do not immediately recognise the mathematical significance by a simple instructions. Suppose, a layer of the given solution is one unit (say 1 c.m.) in thickness or optical path transmits. 50 of the radiant power incident upon it and this radiation becomes incident upon another unit layer of the same material.*

The second layer will transmit 50 of the radiant power incident upon it or (.50 of the .50) .50 x .50 = .25 of the original radiant power and so on for additional unit layers

Optical path - 0, 1, 2, 3, ........... n

Transmittance - 1.0, 0.5, 0.25, 0.125 ..... $.50^n$

The corresponding Figure 10.1 is a diagramatic representation of these facts. The illustration show the exponential or logarithmic relation between transmittance and optical path way length. Mathematically, it can be expressed as-

$$- \log T = a.b$$

Where,

$$T = \text{Transmittance } \frac{P}{P_0}$$

a = absorptivity of the medium

b = optical path length (thickness)

If a solution of unit concentration is assumed, a becomes the absorbance A unit optical path. The negative sign is required in the T term, because transmittance decreases as optical path increases. The term absorbance A used above is the negative log to the base 10 of the transmittance or its identity. The log to the base 10 of the reciptocal or transmittance -

$$\begin{aligned} A &= a.b \\ &= -\log_{10} T \\ &= -\log_{10} \frac{(P)}{P_0} \\ &= \log_{10} \frac{(P)}{P_0} \\ &= \log 10 \frac{(1)}{T} \end{aligned}$$

Because a log to the base 10 is a power of 10, that gives the quantity, we may write also

$$P' = P_0.10^{-a.b}$$

or $$P_0 = P.10^{a.b}$$

All these mathematical expressions are defination of Bouger's law.

## 2. Beer's law

This law also expressed the same relation between transmittance and concentration of absorbing material as Bouger's law expresses between transmittance and optical path. The Beer's law is -

***"the transmittance decreases in geometrical progression as the concentration increases in arithmetic progression"***

Hence,

$$-\log T = a.c$$

Where,

c = concentration

a = absorptivity

and also the absorbance for unit concentration at unit optical path length (Because only log to the base 10 are to be used. The subscript of the base will be omitted hereafter).

For Beer's law, the following relation exists-

$$A = a.c$$

$$= -\log T$$

$$= -\log \frac{(P)}{P_0}$$

$$= \log \frac{(P)}{P_0}$$

$$= \log \frac{(1)}{T}$$

and $P = P_{0} \cdot 10^{-a.c}$

or $P_0 = P_{10}^{a.c}$

## 3. The fundamental law of spectrophotometer

The fundamental law of spectrophotometer is obtained by combining Bouger's law with Beer's law to give following relation:-

$$A = a.b.c$$

$$= -\log T$$

$$= -\log \frac{(P)}{P_0}$$

$$= \log \frac{(P_0)}{P}$$

$$= \log \frac{(1)}{T}$$

and $P = P_0 . 10^{-a.b.c.}$

or $P_0 = P_{10}.a.b.c.$

The forms of these equation shows that a graph of absorbance A (of the given substances at constant optical path) against concentration C is straight line of slope 'a', which is the absorptivity of the substance. Also, a plot of T against 'C' is a straight line of a slop 'a'. Such plot are used as a test for conformity of a system to Beer's law or as a test for deiation from the law.

Strictly speaking, there are no deviation on conformities to the law, when only a straight absorbing species of a chemical compound is present at all concentrations, measured during analysis. Rather any apparent deviation from the law is an indication that at different concentration, these is some change in the nature of the absorber such as association or dissociation, that changes the absorptivity of the solution, being measured.

## VISIBLE RANGE SPECTROPHOTOMETER

### Instruments

The instruments which measure the absorption or emission of the radiant energy from the substances are differently named as following :-

(a) **Photometer**

(b) **Spectrometer and**

(c) **Spectrophotometer**

A brief description of each of them are as follows:-

(a) ***Photometers***

These instruments are used for measuring the fiction of the radiant power at selected wavelength within the spectral range.

The instruments are so designed that there is an entrance slit, a dispersing device and one or more exist slits and a measuring device to measure selected wavelengths within the spectral range.

(b) ***Spectrometers and spectrophotometers***

Spectrometer is an advanced instrument, as compared to photometer. However, the spectrophotometer is the most advanced instrument, which is the modification of the spectrometer. Some extra equipments are associated in the spectrophotometer, too, which helps in the detection of the ratio or a function of the ratio of the radiant power of two electromagnetic beams as a function of spectral wavelength. The two electromagnetic beams may be separated in space and time of both.

## BASIC COMPONENTS

In General, a spectrometer contains the following basic components-

### Source of radiation

I. ***Incident light source***

The radiation source is used for visible and near infrared regions in the form of tungsten or tungsten-iodide incandescent lamp (i.e. electric bulbs etc.).

II. ***Luminious gas source***

This radiant ion source is used for ultraviolet regions in the form of hydrogen or deuterium discharge lamps, operated under low pressure and D.C. connections. They provide continuous UV emission, down to 1650Å with an optical window of fused silica.

## III. Radiation dispersion devices

The dispersing device may be in the following expensive instrument are provided with glass filters or interference filters, which isolate the radiant energy in the system.

a *Glass filters*

They are coloured with pigments and have a band pass range of 350-500 Å at one half of maximum transmittance. Composite filters can be constructed from set of filters (red-yellow series or blue-green series).

b. *Interference filters*

These employ metallic or dielectro layers which produce interference phenomenon at desired wave lengths and rejects unwanted radiation by selective reflection. These filters have a band pass range of 100-150 Å

c. *Prisms*

Better isolation of the radiant energy is done by prism of different qualities. The action of a prism depends on the reflection of light by the prism material. The dispersion, optical density, the beam depend on the variation of the index of reflection with wavelength.

d. *Monochromators*

The prism is a main unit of the chromator systems. The complete unit of dispersing device in a monochromator, consists of entrance and exist slits, suitable bufflers and mirrors.

e. *Slit and Irises*

The proposed of entrance slit is to permit only a narrow source of light, so that after dispersion and refocussing in the plane of exist slit, the amount of overlapping of the monochromatic images is limited.

The exit siit serves to purpose of passing only a narrow band of dispersed spectrum to be observed by the detector. The slits may be provided with the adjustable diaphragms.

An adjustable slit permits only the desired band of radiation.

**IV. Sample holder**

The beam falls on the sample which absorbs a portion of a light and the remainder is transmitted through the sample.

**V. Photodetector**

It is either in the form of barrier layer photocell (photoemission), vacuum tube or an electron multiplier phototube.

The transmitted light through the sample strikes the photodetector, where it is charged into a extremely small electrical signal.

**VI. Amplifier**

The electrical signal is transmitted to the amplifier, where it is highly amplified.

**VII. Recorder or meter**

It indicate the amount of light transmitted through the sample. It is graduated in either linear or curvicidal scale and divisions from 0 to 100 percent transmittance

## SINGLE BEAM SPECTROPHOTOMETER

A single beam spectrophotometer is usually operated at a fixed wavelength and is primarily employed for the quantitative determination of the concentratior of a single component, when a larger number of samples are to be analysed. (Fig. 10.1)

The basic requirement of a single beam spectrophotometer is the high degree of stability, both of the light source and the detector system. The range is usually from 340-650 millimicron with a blue sensitive tube and can be extended to 950 mm by the addition of a red blocking filter and substitution of a red sensitive phototube.

## A DOUBLE-BEAM SPECTROPHOTOMTER

(a) ***Double beam in time***

This instrument is well suited for qualitative analysis, where complex curves are obtained over a large spectral range (Fig. 10.2).

The beam, after passing through the exist slit, is altered between reference and sample compartments, with the help of a complex mechanism of troidal mirror and rotating mirror chopper. Thus, the beam alternatively strikes a photocell (detector) at one time as it passes through the sample and the other time, it passes through the reference.

The output of the detector is an alternating signal, whose amplitude is proportional to the differences in intensities in these two paths.

(b) ***Doubles beam in space***

In this spectrophotometer, two separate light patterns are created by a beam splitter & mirror after the beam has crossed to exist slit (Fig. 10.3).

One of the splitted beam is reflected by mirrors through samples compartment and the other through the reference compartment. Unobserved radiations fall on individual detector. The output is based on measurement of the ratio or intensities in the two channels.

The chopping arrangement for light beam is done at the source side, depending upon the specific problems and suitability etc.

The individual signals can be added, substracted or treated in any manner desired.

## AREAS OF APPLICATION

(a) Estimation of substances in solution.

(b) Determination of stichiometry.

(c) Determination of stability.

(d) Determination of special properties of the substances.

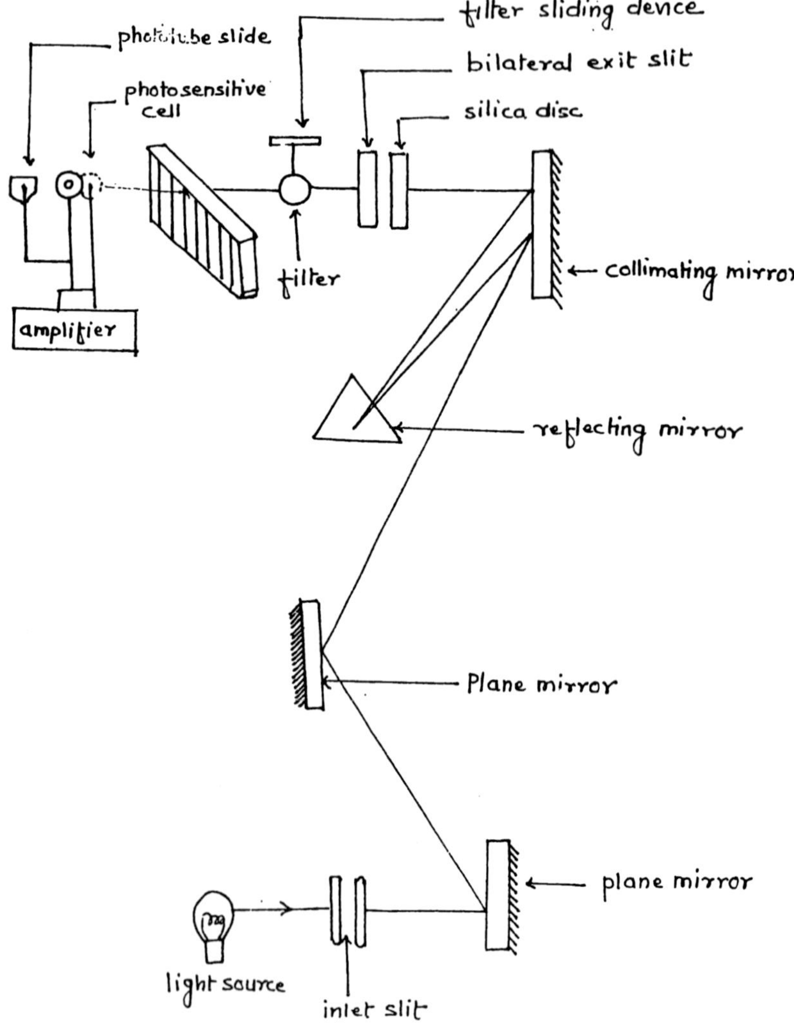


**Fig. 10.1 : Single beam spectrophotometer - schematic arrangement of the parts in a single beam spectophotometer**

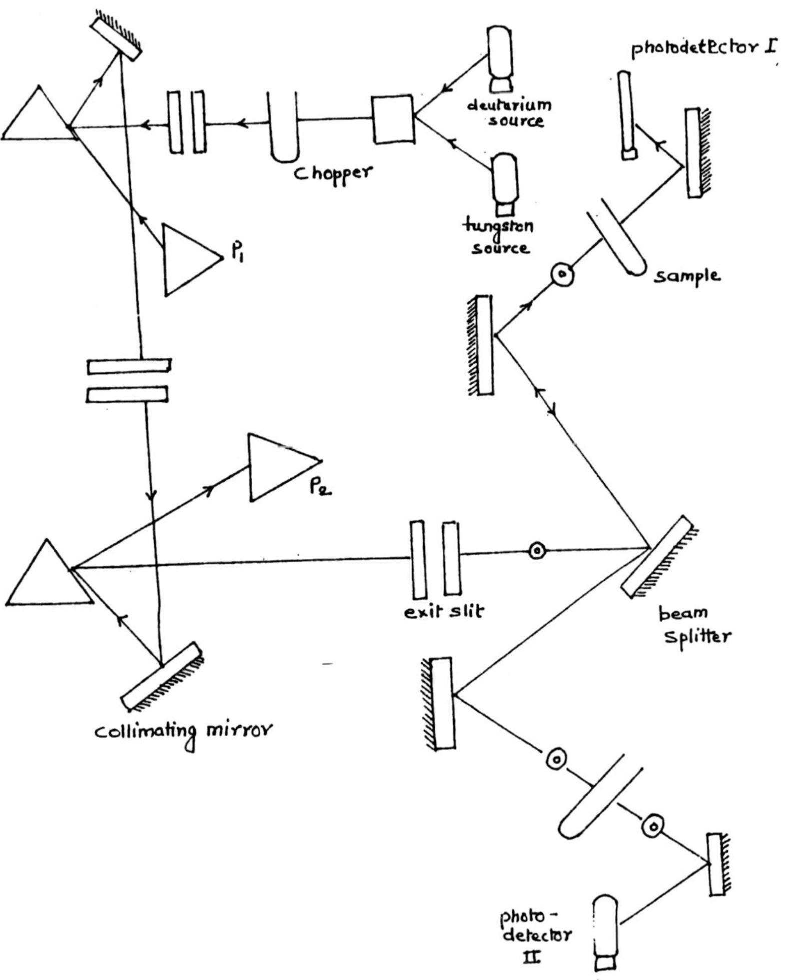


**Fig. 10.2 : Schematic arrangement of double beam in space spectrophotometer**

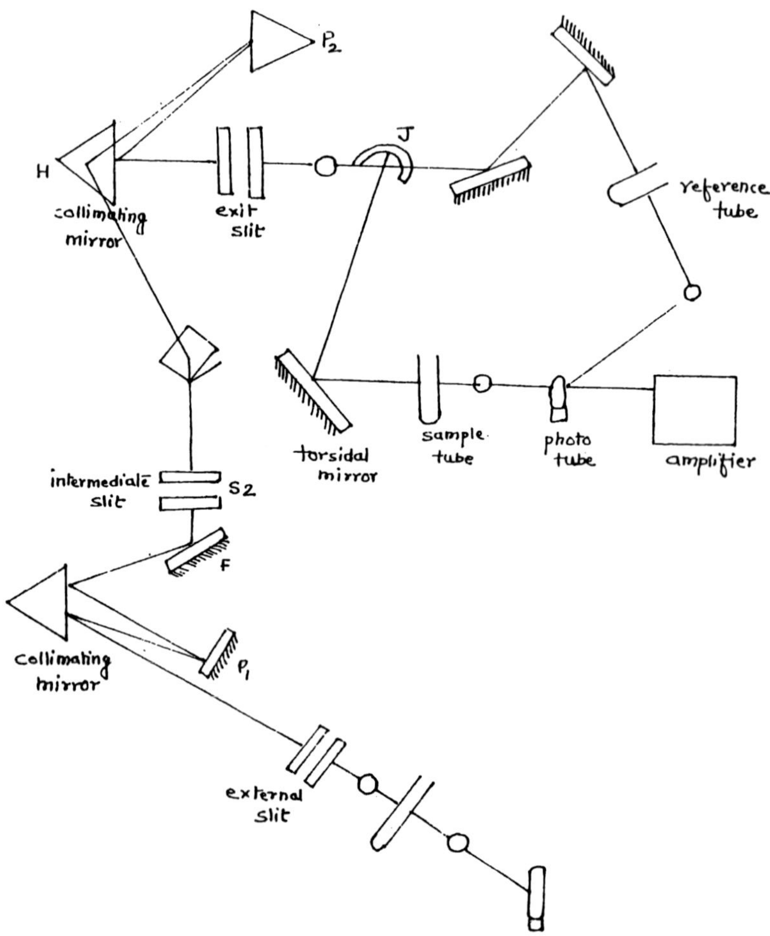


**Fig. 10.3 : Schematic arrangement of a double beam in time spectrophotometer**

# Chapter -11

# ATOMIZATION (ATOM FORMATION) - A CONCEPT

## INTRODUCTION

An aliquot (upto 100 μl) of the sample solution is placed in contact with the inner wall of the cylindrical graphite tube. The tube is then heated to temperature as high as 2000°C in an inert atmosphere to bring about atomization of the sample in order to make atomic absorption atmosphere.

For most elements, metal solutions of oxy-anion salts such as sulphate and nitrate are preferred in the furnace atomizer rather than the chloride solutions normally used in flame atomization. Chloride solutions give a higher degree of molecular volatilization, which decreased the degree of atomization in the furnace.

Most Oxy-anion salts of metals decompose to the corresponding metal oxide on heating. A possible model for the formation of metal atoms would, therefore, be the reduction of the metal oxide by the graphite yielding carbon mono-oxide and free metal atom as per reaction:

$$MO\ (S) + C\ (S) \longrightarrow CO\ (g) + M(g)$$

Where M is the metal being atomized, S and g denotes the solid and gaseous forms respectively.

## INTERFERENCES

The phenomenon that cause deviation in the theoretical atomic spectrometric growth curve is interference. These are classified in the following categories :

1. Spectral,
2. Physical (non specific) and,
3. Chemical (specific)

## 1. Spectral Interferences

It is concerned with radiation emission, absorption and detection without specific reference to the analyte. Spectral interference in silver-copper hollow cathode lamp occurs which may be over come by sequential atomization.

Spectral interference of lead and antimony determination at 217.56 nm, Sb and Pb determination at 217.00 nm and Ni in the determination of Sb at 231.5 nm are similarly compensated. Direct absorption overlaps are useful for analysis of transition metals.

***Background absorption***

Non-specific (molecular) absorption effects are more serious. When a sample is atomized, it may absorb or scatter source radiation because of the simultaneous volatalization of the sample matrix.

The volatilized sample matrix may be in the form of gaseous molecules, salt particles, smoke or other phenomenon. The combined effect is known as back-ground absorption. This generally becomes more pronounced at shorter wave lengths.

Deuterium lamp is used as a background corrector. Radiation from the analyte source lamp and from deuterium are lamp are passed alternatively through the atomizer. The element being determined effectively absorb only light from the hollow cathode lamp, while background absorption effects both beams equally.

When the ratio of the two beams is taken electronically, the effect of background absorption is eliminated.

The other background corrector systems are the :

1. Zeeman background corrector system and,
2. the Smith hiefje background corrector system.

***Zeeman background corrector system***

In this system either the light source or the atomizer cell is placed between the poles of a strong magnet. The magnetic fields splits the emission line or absorption profile into 3 or more components.

The central II component occurs at the original wavelength but is polarized paralleled to the direction of magnetic field, the sigma (6) side band are displaced either side of the original wavelength and are polarized at right angles to the field direction.

A rotating polarizer is therefore used between the atomizer and the source. The parallel light is absorbed at right angles is absorbed the background only correction is obtained by substracting the two signals.

***Smith-Hieftje system***

In this system, the hollow cathode lamp is rapidly pulsed at a high current or background measurement and then at normal current which measures the total signal. Background correction is achieved by substracting the two measurements.

Reducing the sample size placed in the furnace often reduces the background absorption to a greater degree than the analyte signal.

Increasing the purge gas flow will often decrease the residence time of the background producing components. The use of lamp atomization can often be utilized to physically separate the signals produced by the analyte and background absorption producing components.

The use of an alternative analytical absorption line at a longer wavelength will usually give a lessening of background absorption effects. Modification of the sample by chemical means will frequently reduce the effect of background absorption. This technique is known as **'Matrix Modification'**

Matrix modification can be done either directly inside the furnace or outside the furnace. The objective of any matrix modification is to reduce the background absorption interference. The modification can be of two types :

1. Modification that reduces the volatility of the analyte.

2. Modification that increases the volatility of the matrix. In both the cases, the matrix can be removed prior to atomization. However, analyte atomization before matrix volatilization is also sometimes used by using lamp atomization.

## 2. CHEMICAL INTERFERENCES

Also known as ***condensed- phase interference***. The element of interest can combine with other cation or anion in the sample matrix to form a compound which influences the degree of atomization.

As the interference is normally not present in the standard solutions used for comparison, the analytical results will in be in error.

## 3. MATRIX INTERFERENCE

This is also known as ***bulk interference***. These occur, when the physical and chemical characteristics of the sample and standards differ considerably causing a differences in the rate of elemental volatilization during atomization.

Thus for a given amount of element in a sample and in a complex matrix, it is likely that two different atomization signals will be obtained.

Matrix interferences rapidly occur in sample solutions containing high concentration of dissolved salts or acids.

The problem of matrix interference can be overcome in following way :

1. Matching the concentration of major constituents in standard and samples.
2. Method of standard additions, the added analyte should behave similarly to the analyte in sample.

Chemical and matrix interference cause a change in peak height measurements relative to pure standards, the identification of the specific interference (chemical or matrix) is difficult.

Gases generated during atomization may have molecular absorption bands encompassing the analytical wavelength e.g. in case of Cr at 357.9, there is a cynogen absorption. Cynogen band (335-359, 374-389, 410-422) use 360.2 nm for Cr.

**Determination of Mercury**

The mercury in the sample is oxidized to mercuric ion with potassium permagnate in a nitric acid - sulfuric acid medium.

Hydroxylamine hydrochloride is then added to reduce the mercury to metallic form. Then the mercury is vaporized and circulated by the aerator system.

Measurement is made with mercury analyser. The 253.7 nm mercury line emitted by a mercury lamp is absorbed by the mercury vapour in the flow through absorption cell. The change in transmittance is detected by the phototube and displayed on the meter directly in μg mercury. Mercury in the sub microgram range with a sensitivity of 0.01 μg can be determined using Perkin Elmer Model MAS 50A Mercury analyser system. Some reported interferences are presented in Table 11.1.

**Table 11.1 : Interfering elements in the determination of some analytes by flameless atomic absorption spectrophotometry**

| *S.No.* | *Analyte* | *Interference* |
|---|---|---|
| 1. | Cd | Na, Ca, Cu, $ZnCl_2$, Mg |
| 2. | Cr | Al, Fe, Mg, Cu, $H_2SO_4$, $H_3PO_4$, Na, K, Ca, $Cl^-$, $F^-$, Ti, Co, Ni, V, Mn, Sh, Cd, Pb, Ag. |
| 3. | Mn | High Chlorides, $H_2SO_4$ |
| 4. | Pb | $SO_4^{--}$, $Cl^-$, Al, Ca, Fe, K, Mg, $No_3^-$, Ti, P |
| 5. | Cu | $H_2SO_4$ |
| 6. | As | $Cl^-$, S |
| 7. | Al | Si, Sn |
| 8. | Co | $Cl^-$, S, P |

## PREPARATION OF WATER AND WASTEWATER SAMPLE FOR ATOMIZATION STUDY

### Sample handling

Before collecting a sample, decide on the type of data desired i.e. dissolved, suspended, total extractable metals. This decision will determine whether the sample is to acidified, with or without filtration and the kind of digestion required.

Acidify all samples at the time of collection to keep the metals in solution and to minimise the adsorption on the container walls. If only dissolved metals are to be measured, filter the sample through a 0.45 μm membrane filter before acidification.

If possible, filter and aciditfy in the field at the time of collection, report the results obtained on this sample as 'dissolved'.

Filtration is not necessary when total or extractable concentration is required. Acidity the sample with concentrated nitric acid to a pH of 2.0 or less. Usually 1.5 ml. $HNO_3$/lt. sample will be sufficient for potable waters free from particulate matter. Such samples can be analysed with no further treatment. However, sample containing suspended materials or organic matter, e.g. wastewater and raw water samples require pre treatment.

Pretreatment may include additional acid to preserve the sample and certainly requires a digestion step to destroy organic matter and bring all metals into solution.

### Total metal analysis

Transfer the representative portion of well mixed sample (50 to 100 ml.) to a beaker and add 5 ml. conc. $HNO_3$.

Place the beaker on a hot plate and evaporate to nearly dryness, making certain that the sample does not boil.

Cool the beaker and add another 5 ml. acid.

Cover the beaker with watch glass and return to the hot plate. Increase the temperature until a gentle refluxing action occurs. Continue heating, adding additional acid as necessary until digestion is complete, this is indicated by a light coloured residue.

Add 1 to 5 ml. concentrated nitric acid and warm the beaker slightly to dissolve the residue.

Wash down the beaker wall and watch glass with distilled water and filter the sample to remove silicate and other insoluble material.

Adjust the volume from 50 to 100 ml. or more other pre determined volume based on expected metal concentrations.

The sample is now ready for analysis, Report result as **'TOTAL'**.

### 3. Suspended metal analysis

If the concentration of metals in the suspended material is to be measured, use the same digestion procedure after filtering the sample through 0.45 μm membrane filter.

Digest the filter as well as material on it and run a filter blank to permit a blank collection.

Report the suspended metals as micrograms or milligrams per litre or weight of residue and repot as microgram or milligrams per gram.

Report the results as **'SUSPENDED'**

### 4. Extractable metals analysis

Extractable metals include metals in solution plus metals lightly adsorbed on the suspended material. The results obtained in analysis for extractable metals will be influenced by the kind of acid or acids used in the digestion, the concentration of acid and the heating time.

Unless the conditions are controlled rigidly, results will be meaningless and unreproduciable.

The following procedure determines metals soluble in hot HCl-$HNO_3$.

At the time of collection, acidify the entire sample with 5 ml. conc. $HNO_3$/litre sample. At the time of analysis, mix the sample, transfer a 100 ml. portion to a beaker or flask and add 5 ml. 1 +1 redistilled HCl.

Heat 15 minutes on a steam bath. Filter and adjust the volume to 100 ml.

The sample is then ready for analysis.

The data approximate the total metals in the sample although something less that the actual total is measured.

Concentration of metals found, especially in heavily silted sample, will be substantially higher than results obtained on only the soluble fraction.

Report as **'EXTRACTABLE METALS'**

## 5. Determination of Cd, Ca, Cr, Co, Cu, Fe, Pb, Mg, Mn, Ni, Ag, Zn by direct aspiration into air-acetylene flame

***For Mn***

Magnese may exist in soluble form in a natural water when first collected by it oxidized readily to a higher oxidation state and precipitates from solution or becomes absorbed on the walls of containers.

Determine magnese very soon after Collection. When delay is unavoidable, total magnese can be determined if the sample is acidified at the time of collection.

**For Zn**

Analyse sample within six hours after collection. The addition of HCl will preserve the metallic ion content but requires that :

a. the acid be zinc free

b. the sample bottle be rinsed with acid before use and

c. the sample be evaporated to dryness in silica dishes to remove the excess HCI before analysis.

**For Cu**

Copper ion tends to absorb on the surface of the metal container.

Therefore, analyse samples as soon as possible after collection.

If the storage is necessary, use 0.5 ml. 1+1 HCl/100 ml. of sample to prevent 'plating out'.

**For Ca and Mg**

Mix 100 ml of standard and sample as well with 25 ml. of Lanthanum solution before aspiration.

**For Fe and Mn**

Mix 100 ml. of standard and sample as well with 25 ml. calcium solution before aspiration.

**Determination of low concentration of Cd, Cr, Pb by chelation with APDC and extraction into MIBK and aspiration into air acetylene flame**

1. Adjust 100 ml. of standard, 100 ml. of sample and 100 ml. of deionised distilled water blank each to pH 2.2 to 2.8 by adding 1 N $HNO_3$ or 1 N NaOH.
2. Transfer each standard solution, sample and blank to an individual 250 ml. separating funnel and add 1 ml. ADPC solution and shake to mix.
3. Add 10 ml. MIBK and shake vigorously for 30 seconds.
4. Let the contents of each separating funnel separate into aqueous and organic layers.
5. Drain off aqueous layer and discard. Make sure that none of the aqueous layer remains in the stem of funnel.
6. Drain the organic layer into 10 ml. glass stoppered graduated cylinder.
7. Aspirate the organic extracts directly into the flame (zeroing the unit of MIBK blank).

Table 11.2 shows the pH ranges for individual element extraction to obtain optimum extraction efficiency.

It is important to note that the above extraction procedure will measure only hexavalent chromium. To determine total chromium, oxidise trivalent to hexavalent chromium by bringing the sample to a boil and adding $KMnO_4$ solution drop wise to give a persistent pink colour while the solution is boiled for lominutes, cool and is ready for analysis.

**Table 11.2 : pH ranges for individual element extraction to obtain optimum extraction efficiency by ADPC-MIBK**

| *S.No.* | *Element* | *pH range for optimum extraction* |
|---|---|---|
| 1. | Ag | 3-5 (complex unstable) |
| 2. | Cd | 1-6 |
| 3. | Co | 2-10 |
| 4. | Cr | 3-9 |
| 5. | Cu | 0.1-8 |
| 6. | Fe | 2-5 |
| 7. | Mn | 2-4 (complex unstable) |
| 8. | Ni | 2-4 |
| 9. | Pb | 0.1-6 |
| 10. | Zn | 2-6 |

**Determination of Al, Ba, Be, V and Si by direct aspiration into a $N_2O$-acetylene flame**

For Ba, add 2 ml. $CaCl_2$ solution to 100 ml. standard blank and samples each before atomization.

Determination of low concentrations of Al and Be by chelation with 8-hydroxyguinolene and extraction into MIBK and aspirating into $N_2O$-actylene :

1. Take 100 ml. of standard samples and water blank to different separating funnels.
2. Add 2ml. 8-hydroxyquionolene solution to each and shake.

3. Add 10 ml. Ammonium hydroxide - Ammonium acetate buffer solution to each and shake.
4. Add 10 ml. MIBK to each and shake vigorously to mix.
5. Let the content of each separating funnel separate into aqueous and organic layers.
6. Discard the aqueous layer by draining it off.
7. Drain the organic layer into 10 ml. glass stoppered graduated cylinder.
8. Aspirate organic extract directly into the flame (zeroing the instrument on the bank.

## TRACE METALS ANALYSIS (AIR-BORNE MATERIALS) BY AAS

### Collection of air-borne materials

Ambient atmospheric particulate matter and industrial dust and fumes are sampled with cellulose membrane filter of 0.8 μm average pore size.

The flow are 1.5 litre/minute ambient temperature, barometric pressure are recorded at the beginning and at the end of the sample collection period.

A minimum 60 litres should be collected.

### Sample preparation

The samples including clean filter and blanks (one filter and cover blank for every 10 lt. sample) are transferred to clean 125 ml. Phillips or Griffin beakers.

Add conc. $HNO_3$ to cover the sample with watch glass.

Heat on a hot plate at 140°C in a flame hood until the sample dissolves the slightly yellow solution is obtained.

About 30 minutes will be sufficient for the most air sample. However, more $HNO_3$ may be needed to complete ash and destroy high concentration of organic matter and under these conditions, longer time for ashing will be required.

Once the ashing is complete indicated by clear solution, the watchglass is removed and the sample is allowed to evaporate to dryness.

If the residue in the beaker is light whitish material. The beaker is cooled and 1 ml. of $HNO_3$ and 2-3 ml. of distilled water is added.

The beaker is placed on the hot plate and swirled occasionally until the residue is dissolved. The beaker is then removed and solution is transferred with distilled water to a 10 ml. volumetric flask.

**Wet ashing**

Take representative sample of solid waste grind into 20 to 40 mesh size.,

Take 5 gms. of flash i.e. kjeldahl flask, add 20 ml. concentrated nitric acid and 10 ml. concentrated prchloric acid ($HClO_4$).

Heat gently (first in sand bath) and then dissolve the sample ash in a minimum amount of concentrated HCl and warm distilled water. Filter the diluted sample, neutralize with concentrated $NH_4OH$ and adjust to a known volume.

# Chapter -12

# ATOMIC ABSORPTION SPECTROPHOTOMETRIC MEASUREMENT

## INTRODUCTION

The phenomenon of atomic absorption was probably first observed by Wallastone (1802) who repeated Newton's experiment of analysing continuous solar spectrum & revealed that if a beam of light from the sun passes & slit, the solar spectrum is intersected by several darklines.

These lines were later on rediscovered by 'Fraunhofer' & were called '**FRAUNHOFER LINES**'.

Krichoff (1860) for the first time established the presence of certain elements in the sun from the fact that the emission lines for these elements coincides with the Fraunhofer lines in the solar spectrum. Later, it was demonstrated that atomic spectra could be used in emission or absorption on the basic of a new method of analysis.

The atomic absorption in environmental analysis was used for the first time by Woodson for the determination of mercury vapour in the air.

The first atomic absorption spectrophotometer was designed & fabricated by Allan Walsh of CSIRO Australia. The first public exhibition of working AAS was made in March 1954 in Melbourne University.

The first paper of atomic absorption spectro photometry was published by Walsh in 1955 & first instrument was marketed in 1962 by M/s. Techtron.

## PRINCIPLE

Atomic Absorption is based on the absorption of photon by an atomic vapour in ground state. The element of interest in the sample is dissociated from its chemical bands & placed into an unexcited, unionised **'GROUND STATE'**.

This dissociation is achieved by supplying thermal energy to the sample. The element is then capable of absorbing radiation at discrete lines of narrow bandwidth. The wavelength at which absorption occurs is characteristics of the element & the degree of absorption is a function of concentration of atoms in the vapour.

## INSTRUMENT

The basic component of an atomic absorption spectrophotometer are show in Fig. 12.1. Line radiation from a suitable source is passed through an atomic vapour by means of suitable optics to a wavelength selector. The later isolates the wavelength which can be aborbed by the test material from other spectral lines emitted by the source.

A photo sensitive detector measures the light flux later after passage through the wavelength selector. Usually, the output signal required additional amplification before presentation to a meter or recorder. The spectral source is usually modulated to separate the source radiation from that coming from the excited atoms in the atomic vapour.

### Lamp Source

Hallow cathode lamp is the source of line radiation. It consists of a cylindrical glass tube with an end window of borosilicate glass, quartz or fused silica.

In the centre of the tube, lamp a cup shaped cathode which is made from either the pure form of the metal to be measured or an alloy of it when the pure metal is non-conductive or voltive.

The tube is evacuated & filled an ultra-pure monoatomic gas like argon or neon at low pressure. When an electric field is applied between the cathode & anode, a bright discharge appears at the cathode entering a narrow beam of characteristic line of the element. It should radiate the characteristic resonance lines of the element to be determined without interference from other spectral lines originating from impurity elements, electrode material or carrier gas.

Separate hallow cathode lamp is used for every element to be analysed. However, multielement hallow cathode maples are also available. High intensity hallow cathode lamps and electrode discharge lamps are not being used for achieving better sensitivities of detection.

## ATOMIC VAPOURS PRODUCTION IN THE SYSTEM

Atomic vapours are produced in an atomiser. Thermal energy in the form of flame or electrical heat is supplied in order to provide free atomic environment. Therefore, there are two types of atomizers flame & non flame. In flame type atomizer, the pretreated sample is aspirated into air-acetylene flame or nitrous oxide-acetylene flame depending on the element to be determined. Air-acetylene flame gives a temperature of about 2200°C & is suitable for the determination of metals like Ag, Cd, Cu, Cr, Co, Fe, Mn, Ni, Pb & Zn.

Nitrous oxide - acetylene flame gives higher temperature of about 2750°C & is suitable for the determination of refractory metals like Al, B, Zr, Mo, Si, S etc. However, nitrous-oxide-acelylene flames requires a specialised & a more careful handling.

Flaming are the most popular means for producing an atomic vapour from the test material introduced usually as a liquid aerosol. The flame must be able to dissociate refractory molecule species & convert the analyte into its constituent atoms.

When a flame is used, some elements are partially atomized, thus, resulting in loss in sensitivity. The possibility of chemical interferences due to variation in degree of atomization of one element with the concentration of other elements, radicals or compounds in the solution also increases.

This type of interference is responsible for series limitations in the flame methods. One of the foremost problems encountered in atomic absorption analysis is that of the formation of the metal oxides within flames. With the use of increasingly higher temperature flame in order to dissociate the metal complexes present within the flames, ionization of analyte atom is promoted.

Various attachments to the flame burner such as microsampler, tantalum boat, Delve's cup assembly are also used for the analysis of microquantities of samples such as blood. In Tantalum boat, an aliquot of sample is pipetted into the metal both. The sample is then evaporated in outer fringes of the flame & finally atomized in the flame centred in the incident beam.

The Delve's cup atomization system utilizes flame activated atomization from a circular nickel reservoir match as in the tantalum boat sample. In this case, however, the vapour passes into a confining, open ended tube aligned along the source beam of atomic absorption spectrophotometer.

## TYPES OF ATOMIZERS :

### NON-FLAME

A number of non flame atomizers available, as listed below :

1. Wet type filament
2. Tantalium boat atomizer
3. Tantalum strip atomizer
4. Carbon rod atomizer
5. Mini Massman furnace
6. Woodriff furnace
7. R.F. furnace and
8. Graphite tube furnace

Graphite tube furnace (heated graphite atomizer) was developed by L'voze, a Russian physical chemist. It is a graphite tube which can accommodate, a much larger sample aliqust upon 100 μl.

Grooved reservoir are available for organic solutions which tend to migrate within the tube. Tube well temperature upto 2700°C may be reached a although the temperature at the centre of the tube will be several hundred degree cooler, this should be sufficient to atomise most of the heavy metals with a degree of efficiency.

An inert gas atmosphere is maintained within the atomization cell in the usual fashion, but the gas stop during atomization increases the sensitivity. A given amount of sample in the furnace gives much higher absorption then it would be in a burner, as the atomization efficiency of the furnace is nearly 100 percent for most of the elements.

The improved sensitivity in the furnace produces detection limits which are as much as 1000 times better than hose obtained with conventional flame systems.

## MONOCHROMATOR

Monochromator is used as wavelength isolator. It consists of an optical train arrangement for rotating, the grating on its axis for adjusting the wavelength, variable slits for adjusting the intensity of the source beam & the radiation detector.

The optical train rejects all the lines of other elements including those of the filler gas in the light sources & prevents the radiation detector from being overloaded with light. The useful wavelength range runs between arsenic line 193.7 mm & the cesium line at 852.1 mm.

## PHOTOMULTIPLIER

Photomultiplier is used as detector. It is a simple phototube with built in dynodes for emmision. Response time is of the order of nano-seconds.

## AMPLIFIERS & READOUT

The amount of the photomultiplier detector is amplified by A.C. amplifiers tuned to the frequency of modulation of the hollow cathode lamp to cut off the background and is finally given to logarithmic amplifier to get the out put directly in terms of absorbance.

To provide facility like autozero, operation in concentration, mode & integration of signal, additional circuit are given after logarithmic amplifier. The signal is then passed to the readout.

The readout can be a meter, a recorder or digital display. Recorder is most commonly employed for measuring the transient absorption signals, produced by the heated graphite atomizers. The signal output can be recorded in terms of absorbance or concentration in peak height or peak mode.

## SENSITIVITY

Sensitivity using heated graphite atomizer is the weight of an element (usually expressed in picograms, $10^{-12}$g), which produces a signal of 0.0044 absorbance units (1% A). The sensitivity with graphite cell is inversely proportional to the square of the inner diameter of the graphite tube, if small amount of sample are analysed.

Pyro-coating is also done on inert material followed by passing an inert gas stream containing methane through the tube at a high temperature. The sensitivity improvement using commercially available pyrocated graphite tube is short lived when the operating temperature are 2700°C or more.

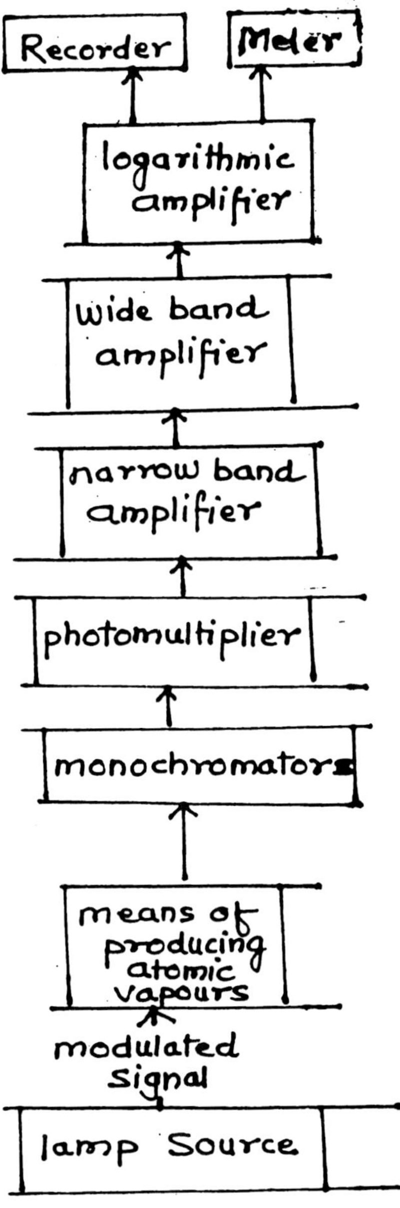


**Fig. 12.1 : Block diagram of the basic Atomic Absorption Spectrophotometer**

## Chapter -13

# ULTRAVIOLET SPECTROSCOPY

## GENERAL INTRODUCTION AND PRINCIPLE

An absorption spectrum is obtained when a substance in solution form or in the gaseous form is placed between the monochromatic light and the detector of the spectrophotometer.

When this energy falls on the molecule, its energy increases as follow:-

$$E = h\nu$$

$$= \frac{hc}{\lambda}$$

Change in the energy can be one of the following form:-

1. Rotational.
2. Vibrational and/or,
3 Electronic.

Rotational changes involve smallest quanta, vibrational relatively larger quanta whereas electronic the largest, as compare to two others. If a small energy is absorbed, then molecules in the atoms pertain to rotational energies and there are in far infrared range (i.e. 25 μ onwards).

If radiation is greater than both vibrational and rotational energies in a molecule energy change, this move occurs in near IR and IR range (i.e. beyond 1 to 25 μ).

If the radiation energy is much larger than changes more in electronic and less in vibrational energies take place as in UV range. If it go to far UV, the changes in electronic energies only take place. Thus in UV range, electronic and vibrational spectra of the substance is studied. In UV spectroscopy, an UV source i.e. hydrogen lamps is used.

## BEER-LAMBERT LAW

The amount of light absorbed is proportional to the molecules present in path. Now, if the substance is in solution, number of molecules state its concentration.

$$\text{Absorbance/ extinction/ OD} = \log_{10} \frac{10}{L} = \text{e.c.l.}$$

Where,

| | |
|---|---|
| e | = absorption coefficient |
| c | = concentration |
| l | = length |

If,

| | |
|---|---|
| c | - in gm moles / lt. |
| l | - in c.m. |

Then ,

l is the ***molar extinction coefficient***

## STANDARDS

For the wavelength calibration purposes in UV, a standard mercury source is kept at the place of lamp and the principle lines of mercury of 257 and 365, 401, 546 are seen.

Secondary standards are the solution of salts :

1. Potassium chromate salt.
2. Copper sulphate.
3. Cobalt ammonium sulphate.

Standard absorption maxima of there can be obtained from literature and the instruments set for it. In UV, solvents are also very important. The important factor of solvents is that the organic solvents under choice should not absorb at the absorption of the substance. Some of the limiting values for solvents are shown in Table 13.1

**Table 13.1 : Limitation values for some solvents**

| *Solvent* | *Limiting values* |
|---|---|
| Benzene ($C_6H_6$) | 280 |
| Carbon tetra chloride ($Ccl_4$) | 265 |
| Chloroform ($CHCl_3$) | 245 |
| Ethanol ($C_2H_5OH$) | 210 |
| Methanol ($CH_3OH$) | 215 |
| Water ($H_2O$) | 210 |
| 1, 4 dioxane | 220 |
| Cyclohexane | 210 |

**Table 13.2 : Chromophonic groups and solvents**

| *S. No.* | *Chromophonic group* | *Group rendering* | *System (example)* | $A_{max}$ | *Solvent* |
|---|---|---|---|---|---|
| 1. | Carbonyl | RRIC=0 | Acetone | 188 | n-hexane |
| 2. | Carboxyl | RCOOH | Acetic acid | 204 | Water |
| 3. | Nitroso | N=0 | Nitroso-butane | 300 | Ether |
| 4. | Nitrate | $-ONO_2$ | Ethyl nitrate | 270 | Dioxane |
| 5. | Nitrite | -ONO | Amyl nitrate | 218 | Petroleum ether |

These are the cut off wavelength substance from 0.1 to 100 mg is sufficient for obtaining a spectra.

## CELLS

UV cells are quarter as glass absorbs the energies in this range. Table 13.2 presents some of the chromophonic groups and solvents.

Some of the certain phenomena occurring in UV for the changes in electronic levels and the electronic spectra can be given as :-

1. ***Chromophores***

   Unsaturated groups responsible for the electronic spectra e.g. C= C. = 0, -ONO etc.

2. ***Auxochromes***

   Saturated groups when attached to chromophores, alter wavelength and intensity of absorption maxima e.g. OH, $NH_2$, Cl etc.

3. ***Bathochromic shift***

   This is on the longer wavelength side due to the change in different solvents.

4. ***Hypsochromic shift***

   Shift in a maxima to shorten wavelength side due to the changes in solvents.

5. ***Hyperchromic shift***

   This is increase in absorption intensity due to changes in the solvents

6. ***Hypochromic shift***

   This is decrease in absorption intensity.

## APPLICATIONS

This is increase in absorption intensity due to the solvents that are used mostly the organic groups or the inorganics associated with organics can be studied in this region.

Chapter -14

# INDUCTIVELY COUPLED PLASMA ATOMIC EMISSION SPECTROMETRY (ICP-AES)

## INTRODUCTION

The development of ICP's began in 1942. Its applicability for the analysis of toxic & trace metals in the environment gained popularity in past decade. Argon supported ICP's operated at atmospheric pressure are excellent vaporization-atomization-excitation-ionization sources for analytical atomic emission spectroscopy.

When a polychromator is used for observing the emitted spectra, the metals and metalloids can be determined simultaneously at the ultra trace, trace, minute & major concentration levels under one set of experimental parameters. Alternatively, programmable scanning spectrometers may be utilized for sequential determination. The atomization-exctation process is remarkably far from inter element interactions, the powers of detection in ppb range for most elements, and sample manipulation requirements prior to analyses are often minimal. The technique meets the requirements of an analytical system for sequential determination of the elements at all concentration levels to an unusual high degree.

## PLASMA

Plasmas are gases having a significant fraction of their atoms or molecules ionized magnetic fields readily interact with plasmas & one of these interaction is an inductively coupling of time varying magnetic fields with the plasma.

## INDUCTIVELY COUPLED PLASMA (METHOD OF GENERATION)

A high frequency current, provided by X-F-generator is directed to an induction coil, thus creating an oscillating magnetic field, whose lines of force are oriented on the coil axis.

The induced axial magnetic fields cause charged particle (electrons & ions) with the coil to flow in closed annular paths. The electrons (and ions) meet resistance to their flow, resulting in the formation of heat that cause additional ionization (Electric field coupling is also involved in the early stage of formation of a plasma of extended dimension whose temperature may reach 6000°K to 10000°K, if argon flows of the proper centrifugation & magnitude pass through the coil, an adequate seed of electrons is initially provided & the output impedence of the generator is properly matched to the plasma.

## TORCH

It consit of a series of 3 concentric quartz tubes & is shown in Fig. 14.1. The typical value of argon flow is 1.5 lit/mt. & that of water for cooling induction coils is 5 li./meter.

## PRINCIPLE OF OPERATION

Radio frequency radiation is used to heat a flow of argon gas into a plasma by means of an induction coil. When a sample is introduced into this extremely hot (6000°K to 10000°K) argon plasma, the sample is broken down into individual atoms, which are then further excited by the plasma.

These excited atoms then remit this energy as electromagnetic radiation (light), which is characteristic of their respective chemical elements.

The emitted light is then passed into the spectrometers. They disperse the light, separating the various emission lights present in the radiation.

### Simultaneous ICP-AES

At appropriate positions, photo multiplier tubes are positioned. These convert the light energy emitted by atoms into electric currents.

The current from each photomultiplier tube is integrated & then digitized, so that the data may be fed into a computer for storage calculation & printing.

## Sequential ICP-AES

The emitted radiations is measured by a sensitive scanning monochromator, which moves rapidly from wavelength to wavelength locating until all $HNO_3$ has been driven off & the solutions begins to char, add more $HNO_3$, heat until white fumes of perchloric acid are $HClO_4$ evolved out & solutions became clear pale yellow in colours.

Cool & add 10 ml. conc. HCl, heat to dissolve all salts finally cool & make 100 ml. volume by deionised distilled water. A residue of sand & other silica material is removed by centrifugation & discarded. This is called ***master solution*** & is diluted to different dilutions.

## DRY ASHING

Evaporate sample of suitable size to dryness on a steam bath in a platinum or other evaporating dish. Transfer the dish to a muffle furnace & heat the sample to a white ash.

The ashing temperature to be used depends on the elements to be determined in the samples. If volatile elements are present, keep the temperature to 400 to 450°C for as many hours as required.

If sodium only is to be determined, ash the analyte peak and integrating the intensity for desired time. These integrated readings are then converted to concentrations by a computer, displayed on the video screen & if desired, on standard line printer etc.

## SAMPLE INTRODUCTION

Samples are most commonly introduced into plasmas as aerosoles generated from aqueous or organic solutions either by pneumatic through peristaltic pump or ultrasonic nebulization techniques.

Fig. 14.2 shows schematic of a typical pneumatic nebulization facility. Samples in this forms are most convenient for the direct examination of liquids, including process streams, edible oils, liquid fuels & body fluids etc.

Solid samples may be dissolved by a variety of methods but the direct sampling of solids without prior dissolution is clearly advantageous. The transformation of solid metal samples into aerosoles that are the subsequently injected into the plasma has been achieved in several ways e.g. aerosols of solid metallic particles can be produced by ultrasonic nebulizaiton of molten metals.

For effective atomization & excitation, the sample aerosols should be injected into the plasma & remain in the interior high temperature environment of the plasma as long as possible. The residence times & temperature experienced by the sample are approximately twice those found in nitrous oxide-acetylene flames, the hottest combustion flames commonly used in AAS.

The temperature measured above the coil region & estimated by extrapolation down into the induction region shown in Fig. 14.3. The combination of high temperature & relatively long interaction time lead to an unusually high degree of atomization of analyte species, especially for the small aerosols particles resulting from solutions with total salt contents of 0.5 to 2 percent by weight. With high degree of atomization, the vaporization-atomization interface effects are unlikely to a serious problem.

## INTER-ELEMENT INTERACTIONS

The inter-element interactions in ICP are small or insignificants under the experimental conditions that provide excellent power of detection. It is, therefore, possible to establish a single set of calibration curves for the detection of analytes in variety of sample matrices, e.g. The observed/ relative intensities of the analyte line of Al, Ba, Cd, Ca, Cu, Li, Mg, Mn, Va & Zn does not show a significant change when the matrix is changed from relatively pure water to a 0.1 percent solution of $NH_4Cl$, KCl, $MgCl_2$, $CaCl_2$, $AlCl_3$, $FeCl_3$ or $(NH_4)_2$ $HPO_4$.

Spectral line interferences are common in ICP-AES & these are well known approaches for eliminating or drastically reducing their effects on analytical results.

Interferences can frequently be eliminated by judicious selection of lines to be used. Although this may require some compromise between optimum detecting power & least interfence from major between contaminants, such compromise are generally acceptable.

When concentration are near the detection limit, the spectral background will normally be a large fraction of the total measured signal & precise background corrections will be required for accurate analysis.

Changes in concentration of contaminants may produce subtle changes in the background level. True background shifts, if not measure accurately & substracted from the total, may introduce an analytical bias that may appear to be an inter-element or matrix effect.

Advantages of ICP-AES over atomic absorption spectrophotometer :

1. An ideal multi element analytical system.
2. Determination of all the elements is possible.
3. Simultaneous multi element. determination from major to ultraviolet levels - it is important to distinguish between multi element & simultaneous multi element determination.
4. Absence of inter elements effects. Internment interaction effects in ICP-AES are small if proper background correction are applied. However, they are far smaller than the matrix effects observed in AAS, XRF & SSMS.
5. Applicability to microlite & microgram samples techniques base on Non flame AAS has a lower absolute detection limit for microlitre volumes of solutions than does ICP-AES.

   However, these techniques are very sensitive to inter-element interferences. If atleast 0.5 to 1 ml. of sample is available, the relative powers of detection of ICP-AES & NFAAS are comparable.
6. Direct analysis of solid, liquids & gases with minimal sample preparation or manipulation.
7. Capability of performing rapid analysis. The time required for the simultaneous detection of 40 or even more elements with commercial instruments is approximately 2-3 minutes for the complete cycle.
8. Accurate precision & accuracy with the application of high quality polychromator readout systems.
9. Commercial availability at acceptable cost.

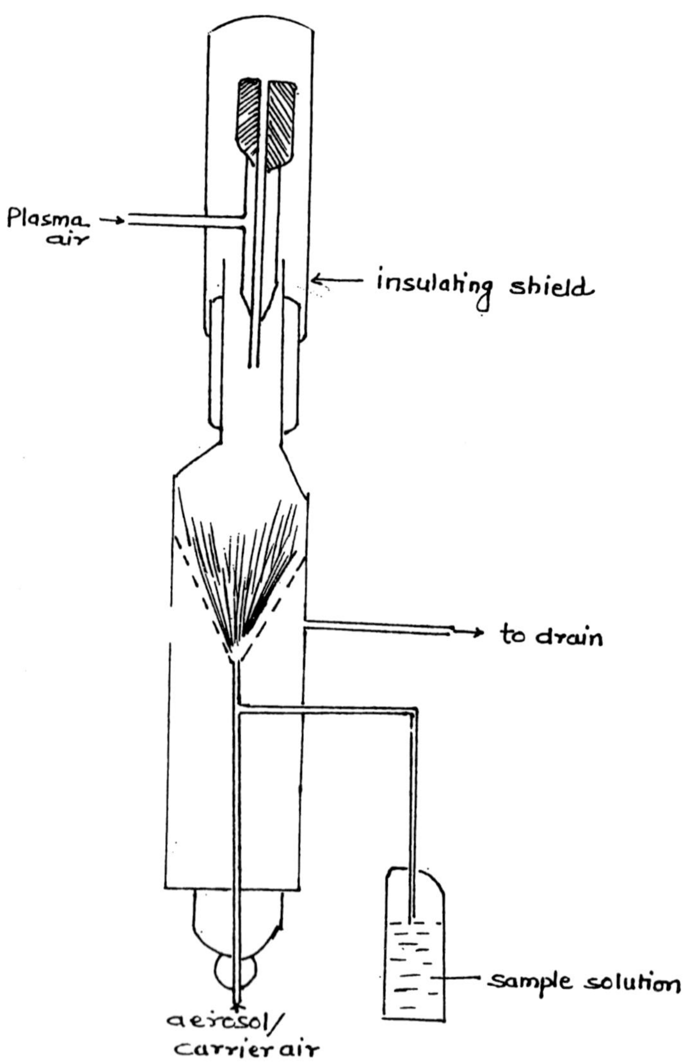


**Fig. 14.1 : A typical nebulization facility**

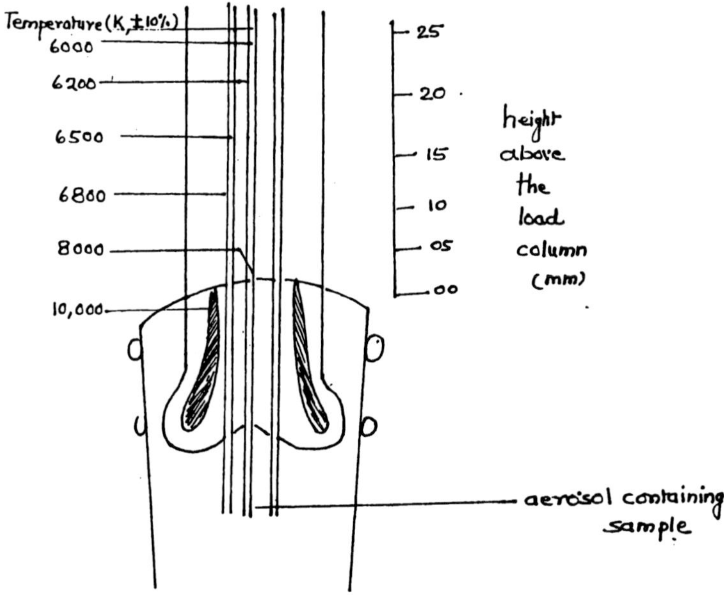


**Fig. 14.2 : Schematic diagram of sample flow through plasma and the temperature(s)**

# Chapter -15

# FLAME PHOTOMETRY

## INTRODUCTION

Flame photometry offers excellent possibilities for the development of analytical procedures for those metallic ions, which can be excited by relatively low excitation levels available in the flame. Interest in quantitative procedures has increased rapidly in the past few years. This has been mostly due to the need for the simpler and more accurate methods for determination of ions of alkali metals.

Special attention has been given to the determination of sodium and potassium because of their importance in biological system. Although flame spectroscopy was first studied by early workers Bunsen and Kirchoff, it was Lundergarh, who studies flame photometry exhaustively and established it as the most important analytical method for the determination of the alkalies.

In this method of analysis, the solution of the sample under test is atomised and finally sprayed into the flame, where the characteristic wavelengths (colours) of the element are imparted and intensity of this colour varies with the concentration of the ions in the flame.

Thus, we see that every flame photometer will consist of following parts :

1. ***Atomiser***

   Here, the sample in the form of solution droplets is broken into fine particles and with the force of the compressed air or oxygen is carried to the flame.

2. Flame and fuel gas where the fine particles of the sample are excited and impart characteristic colour having certain wavelength.

3. ***Measurement of the intensity of the colour***

   For measuring the intensity of the colour, photocell and galvanometer or microammeter are used. The simple layout of the instrument is presented in Figs. 15.1 and 15.2.

   i. In can be observed from the layout that the air at a given pressure is blown into the atomiser and the section thus produced draws the sample, it is broken into fine particles and is carried out with the compressed air to the flame.

   ii. In the mixing chamber, the compressed air meets with the fuel gas at certain pressure and the mixture thus formed is fed into the burner giving out a characteristic flame.

   iii. Radiation from the flame passes thought the lens and the iris diaphragm (which is the mechanical means of varying the amount of radiation falling on the photocell) through a filter which allows the radiation characteristic of the element and falls on the photo cell.

   iv. The optical path from the flame to photocell is enclosed in alight tight box so that no stray radiation reaches the photocell.

   v. The current developed in the photocell is measured by galvanometer.

## PRINCIPLES OF LINES EMITTANCE AND FORMATION OF SPECTRUM

Neils Bohr in 1913 pointed out that when the electrons moves in a closed orbit, it does not emit the energy and this is called as the ***stationery state of the atom***.

The production of the spectral line is attributed to a radiation of energy when transition of an electron from a state of higher energy to that of the lower energy occurs.

Thus, if :

$E_2$ = Energy in the higher energy state of the atom.

$E_1$ = Energy in the lower state of the atom.

Then emission of energy is represented by

$$E_2 - E_1 = h$$

or frequency of emission = $\frac{h}{E_2 - E_1}$

If a substance is exposed to a condition wherein absorption of energy occurs, i.e. by spraying in the higher temperature, the electrons which are usually in the normal state move to the orbits of higher energy.

This state of atom is known as the ***excited state***. Now the electrons have moved to the higher energy orbits which are not their normal orbit. They are in unstable state. As soon as they return to their own orbits, a jump from the higher to the lower state occur with the emission of specific energy or a wavelength of definite frequency.

Thus. we see that the electrons in the outermost orbits are concerned with the formation of the spectra. Since many different energy states are possible in every atom, numerous lines will be observed out of which only principal lines are of great concern. Since they are most intense and are characteristic of the substance and are easily excited.

## FACTORS

### 1. Flame

In the flame photometer, flame is a mixture of the fuel gas and compressed air or oxygen, which is used for the atomization. Different combination of flames are useful for exciting the atom having highest atomic number in the periodic table. Vallee and Bartholomay have worked and collected the data, which is present in Table No. 15.1.

Interesting experiments to raise the temperature of the flame and thus the large number of elements excited were carried out by Vallee and Bartholomay.

**Table 15.1 : Different combination of flame for exciting the atom**

| *Fuel Gas* | *With air at 0°C* | *With $O_2$ ($O^2C$)* |
|---|---|---|
| Hydrogen | 100 | 2700 |
| Acetylene | 2200 | 3100 |
| Methane | 2000 | 2700 |
| Propane | 1900 | 2800 |
| Manufactured Gas | 1800 | 2800 |

They atomised the sample with oxygen and mixed with equal amount of cyanogen and mixture was the taken to burner of 1 mm bore. This flame has a temperature at the centre of about 4600°C. But the possibility of fusing cyanogen is ruled out since it is very toxic and flame produced ultraviolet radiation, harms the eyes.

The flame required should be very steady, quiet and colourless, otherwise the colour of flame may give the erratic results. Since they may occur fluctuations in the flame and this may add to the wrong results. To avoid this, following method is suggested.

Suppose we determined sodium in the given sample. Then prepare two standard solutions of sodium, one which will give galvanometer deflection slight below to that of sample, say $S_2$ and other giving deflection slightly more that of the sample, say $S_L$. We can take three or four sets of reading and write them according to the following Table No. 15.2.

In the last set there seems to be a drift in the flame which effect by about, say six divisions and so reading for the sample can be corrected. Note that this fluctuation in flame occurs due to slight change in pressure of fuel gas and so the fuel and atomising gas pressure should be constant and stable.

**Table 15.2 : Recording of Observation**

| *Set* | *Readings* | | |
|---|---|---|---|
| | *I* | *II* | *III* |
| $S_1$ | 66 | 67 | 73 |
| Sample | 65 | 65 | 71 |
| $S_2$ | 57 | 57 | 63 |

## II. Atomizers

There are two types of atomizers. One right angle type, which is found in Carl Zesis instrument. Here the stream of air is blown across the opening at right angle to the axis of the sample inlet jet. A partial vacuum is produced in the sample inlet tube and liquid is drawn for atomization.

Larger droplets are broken into fine particles and are carried to the flame with the stream of air. The rate of atomisation is found to be 10 ml. per minute. Thus, there is large consumption of sample but only small portions of it reach the flame. This is disadvantageous when little sample is available.

Perkin Elmer flame photometer has got similar types atomize II. Concentric type is shown in Fig. 15.3. The air is forced through outer annuals and creates suction in central tube and sample is drawn to the air stream and fine mist is found. The sample consumption is very less i.e. 1 ml. per minute. The dimensions of orifies is considerably smaller in this case. This is presented in Fig. 15.4.

## III. Optical System

Radiation is collected from the reflector a lense system and passes through filter and falls on photocell. The maximum selectivity is obtained by using the reference filters. The interference filter consists of a thin circular glass base on which are deposited successively by evaporation in high vacuum, a partially transparent film of metal, a film of transparent dielectric and partially transparent film of metal. These layers are guarded by cover glass.

These interference filters have the advantage, in that they can be obtained with peak transmission at particular wavelength according to the requirements. In place of interference fitter, monochromators are used in spectrophotometric attachment.

The type of receptor used for receiving the characteristic radiation and converting it into current photovoltaic cell or (selenium barrier cell) is used for the purpose. The photovoltic cell consists of transparent layer of metal covering insulated from a layer of selenium on a stell backing. Light passes through metal releases electrons from selenium and sets current which is proportional to the incident illumination or intensity.

Here the current is large a hence can be measured directly by sensitive microammeter or galvanometer, but this cannot be electronically amplified since photovoltic cell has low internal resistance. In Beckman and Perkin Elmers instrument, phototube is used. This vacuum type phototube consists of a sensitized cathode which emits electrons when light strikes it and anode held at positive potential with respect to cathode by external battery or power supply.

The electrodes are kept in evacuated bulb and electrons emitted at the cathode are attracted by resistance of phototube is very high and hence electronic circuit for the amplification of current can be built. Thus, the accuracy of these instruments is more.

## PHOTOMULTIPLIER TUBE

This is used where high sensitivity is needed. It has photo cathode and several diodes and then an anode.

## DIFFERENT METHODS OF ANALYSIS

First method of reducing the errors is to prepare the standards for calibration having nearly same composition with respect to unknown.

Special care is required in the preparation of standard solution for the calibration. Also accuracy is influenced by the composition of these solution.

These extraneous element may act upon the intensity of emission of the test the elements. However, if the extraneous containing only test elements can be used, otherwise it is advisable to prepare the standard solutions of approximately the composition of the unknown. This method of calibration is most common and moreover, trial and error method of determination of the test elements is used.

If the extraneous element are present in comparable amounts and the accurate results are needed, the above most simple method becomes somewhat restricted in its uses. In such cases Radiation Buffer method is found to be very useful.

It is a well known fact that radiant energy emission of metal ions is affected by the presence of other cations. some cations improve the emission while some suppress the emission. D.N. Ivanov has studied the effect thoroughly and stated that for flame excitation, the emission intensity of an alkali is increased if another alkali is added.

Because of the extent and the variable nature of the interferences, the previous workers thought if high consistent concentration of diverse cations were added to the sample, the relatively small concentration variations would have no effect on the emission strength of metal ions and developed the so called 'Radiation buffer' method.

This method prevents mutual radiation interferences where it is only required to determine sodium, potassium and calcium, the radiation buffer can be prepared.

Radiation buffer are nothing but the successively saturated solution of all other interfering elements except the test metal solutions. Thus if we concentrate only on the determination of sodium, potassium and calcium, the following will be the type of radiation buffers to be prepared.

## RADIATION BUFFERS :

1. ***For Sodium***

   Distilled water successively saturated with reagent grade chloride of K, Ca and Mg.

2. ***For Potassium***

   Distilled water successively saturated with reagent grade chloride of Na, Ca and Mg.

3. ***For calcium***

Distilled water successively saturated with reagent grade chloride of Na, K and Mg.

Working curves of concentration versus percentage transmittance or galvanometer readings should be prepared for sodium and potassium from the standards containing the proper amount of appropriate radiation buffer. This is usually obtained arbitrarily found to give best results by adding one volume of radiation buffer and 25 volumes of standard or sample under test.

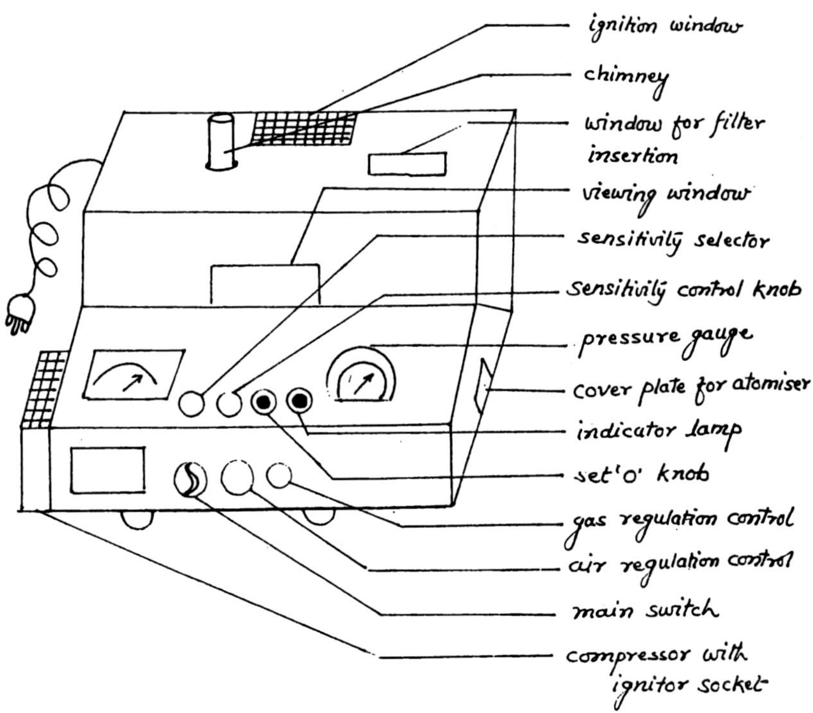


**Fig. 15.1 : Flame Photometer**

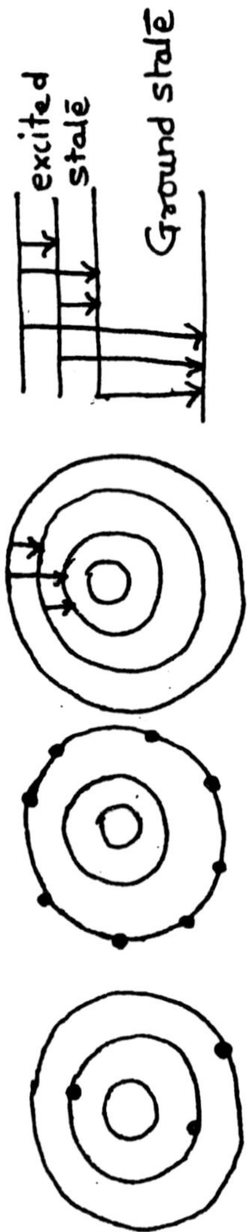


**Fig. 15.2 : Ground and excited state of atom**

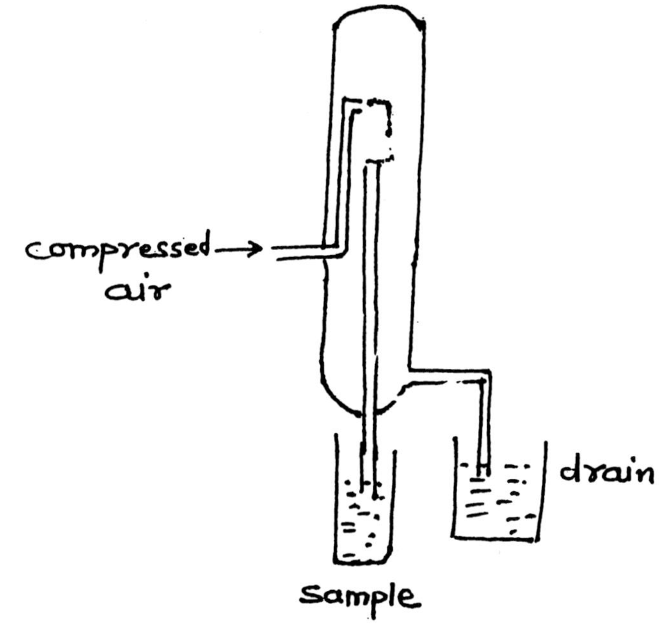


**Fig. 15.3 : Atomiser**

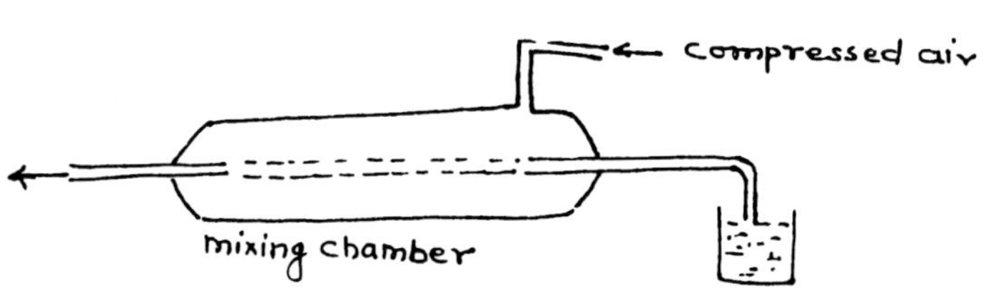


**Fig. 15.4 : Atomiser**

# Chapter -16

# CHROMATOGRAPHY

## INTRODUCTION

The word chromatography was originally applied by the polish botanist **Michael Tswett** in 1906 to a procedure, he applied for separating a mixture of different coloured pigments. He poured the mixture of these pigment dissolved in the petroleum ether through a column of absorbent (calcium carbonate) firmly packed in a narrow glass tube & allowed to drain down the column.

He observed that separate coloured zone of different pigments have been formed in the column. When more solvent (petroleum ether) was added, the zones were separated. Tswett called this column a ***"chromatogram"*** and the process is as "chromatographic method".

## DEVELOPMENT OF CHROMATOGRAPHIC METHOD

Chromatography is a special technique of separating different substances from one another & is useful in quantitative analysis and isolation of various compounds, present in a mixture. At present, chromatographic procedure is also used in assessment of quantitative presence of any substance, present in a mixed aqueous environment. Chromatographic procedure of one kind or another are more widely used in biochemical researches, environmental science and engg., public health engg., and other research & development laboratries.

The important procedures of common use in chromatography are as under -

1. Adsorption chromatography-column chromatography
   a) Liquid solid system column chromatography

b) Liquid system counter current system.

c) Liquid gas system counter gas chromatography.

2. Ion-exchange chromatography (column chromatography)
3. Paper chromatography
4. Sephadex chromatography
5. Thin layer chromatography (TLC)
6. Gas chromatography (GC)
7. High pressure liquid chromatography (HPLC)

## COLUMN CHROMATOGRAPHY

The technique of column chromatography for the isolation of various compound is based on the adsorption capacity of certain substances.

Adsorption is a phenomenon of interaction of the intermolecular forces between the surface atoms of the absorbent and an external molecule to be absorbed. The adsorption may occur due to any of the two phenomenon-

1. ***Physical adsorption***

   If the adsorption is resulted due to the forces created by hydrogen bonds, is known as ***physical adsorption***.

2. ***Chemical adsorption***

   If the adsorption is due to the forces existing between covalency or ionically bonded atoms C adsorption of chemical nature, is known as ***chemical adsorption***. This technique will be delt in ion-exchange chromatography.

The physical adsorption is of great importance, because it is weaker than chemical adsorption and is easily reversible.

## PRINCIPLE OF COLUMN CHROMATOGRAPHY

Tswett (1906) described the difference in the behaviour of pigments placed this column to differences in adsorptivity. The more strongly adsorbed chlorophyll displayed along the column under the influence of the solvent.

As such, the phenomenon of adsorption is involved, which is defined as-

***"The adhesion of a thin layer of molecule by a gas or dissolved substance or a liquid to the surface of a solid body, when a mixture of compounds dissolved in a suitable solvent is allowed to pass through a column of an adsorbent with the help of a specific solvent."***

Usually the same solvent is used, in which, mixture is dissolved. The different compounds are separated and layered one above the other depending on the element capacity (i.e. solvent strength) of the solvent and the differential capacity of the absorbent for the different substances.

The compound occupying the lowest layer in the series of the column indicates the least capacity of being absorbed by the adsorbent, whereas the compound at the top layer shows the greatest adsorbability by the adsorbent of the column.

Therefore, however, instances, which can be explained on some other basis e.g. a mixture of amino acids can be chromatographed on a column of starch or powdered cellulose by the use of certain solvents, but the order of separation depends on their distribution co-efficient between the solvent and water. Since starch or cellulose absorbs sufficient amount of water. It is assumed that the amino acid get dissolved in the water of the column & are extracted from it by the following solvent in proportion to their relative afficinities for the two phases (partition chromatography).

## DISTRIBUTION CO-EFFICIENT

If a substance is soluble in any two solvents, which are immisible and in contact with each other, if it is found that, at equilibrium, the substance is distributed between the two liquid phases in a characteristic manner, the numerical expression of this property is the distribution co-efficient, which is also known as partition co-efficient k, may be defined as -

***The ratio at equilibrium of the concentration $C_a$ of substance in phase A to its concentration $C_b$ in the phase B. It can be expressed mathematically in following form***

$$K = \frac{C_a}{C_b}$$

The value of K is characteristic for a particular set of conditions such as temperature, pressure and presence of other solutes.

## FACTORS INFLUENCING ADSORPTION

The adsorbent capacity of an adsorbent is determined by-

a) its surface area.

b) chemical composition of the exposed surface and,

c) the manners of arrangement of the surface atoms and grains.

On the other hand, the inherent capacity of any compound, to be absorbed, depends on the type aqueous in nature. There are some solvents, which exhibit different nature during the operation of the system. Some of the solvents shows component specific affinity during the separation.

Following solvents are commonly used in the chromatographic technique:-

a) dilute acids

b) Dilute alkalies

c) Alcohols

d) benzene

e) Ether

f) chloroform etc.

## PREPARATION OF A COLUMN

For column chromatography, an appropriate column can be prepared from any of the above mentioned absorbent (s). Column may be very as dimensions are concerned. It is, however, expected that, in general, the column should resemble to the common volumetric buret, as length to weights ratio is concerned. Every column has limitation for separating a partical compound through it, at which, it shows maximum separation capacity. However, if this capacity is exceeded, it diminish the separation.

Column can be regenerated for further use by flowing down all the components by using an appropriate solvent.

## EXPERIMENTATION

### 1. Column preparation

Column can be prepared in following way-

(a) Clean the buret and clamp it on the stand.

(b) the nature of solvent or eluent and,

(c) the temperature.

If there is any change takes place among any of them, the adsorption of a particular compound is get influenced. Consequently, change can be observed either in positive direction or in negative (Fig. 16.1).

## APPLICATION IN ENVIRONMENTAL SCIENCE AND TECHNOLOGY

In Biological sciences, this system applies, quite frequently for isolation/separation of non ionic constituents, present in a mixed aqueous system. e.g. pigmented fractions, fatty acids, lipids etc.

## COMMONLY USED ABSORBENTS

To isolate/ separate out the various constituents of a mixed aqueous system, following absorbents are applied frequently in such chromatographic analytical procedures:-

a) Cellulose

b) Starch

c) Calcium carbonate ($CaCO_3$)

d) Sodium carbonate ($Na_2CO_3$)

e) Magnesium carbonate ($MgCO_3$)

f) Alumina ($Al_2O_3$)

g) Charcoal

h) Magnesium silicate ($MgSiO_3$)

i) Bentonite

j) Ion exchange resin etc.

Various insoluble inorganic oxidizers, phosphates, silicates, sucrose etc., are also in use as common absorbent.

## COMMONLY USED ELUENTS

1. In chromatographic analysis, the elvents or solvents which apply to the system, either aqueous or non-close the stop cork of the buret.
2. Insert some glass wool at the bottom of the buret and then put a layer of sand (approx. 1.5 c.m. thick) over the glass wool.
3. Pour dry alumina into the buret slowly to form a layer of 7.5 c.m. thickness over the sand.
4. Cover the alumina layer with sand (1.5 cm thick).
5. Open the stop cork and wash the column with benzene.
6. Close the stop cork and fill the column with benzene till it comes 2-2.5 c.m. above the upper layer of the sand.

   Now, the column is ready for use.

## PREPARATION OF LEAF EXTRACT

Follow the following steps :

1. Grind 5 gms of green leaves with 20 ml. of mixture of benzene and methanol in 2:1 ratio.
2. Filter the extract through a filter paper and pour the filtrate into a separating funnel. Add 10 ml. of water to remove methanol.
3. Remove the coloured layer of benzene.
4. Concentrate, if necessary, by heating on the hot water beaker (Please don't heat directly).

## CHROMATOGRAPHIC STEPS

1. Pour the concentrated extract gently. Cover the benzene layer of the column.
2. When the extract soaked into the sand, elute it with benzene (flow some benzene in the column). Wait for some time till different coloured bands of pigments are visible.

3. Continue to elute the column with benzene and go on collecting different pigments one after the other, separately in different tubes (by opening the stop cork of the buret).

4. If any colour band is left in the column & is not removed by benzene, then, add some methyl alcohol for eluting the benzene.

5. Examine these fractions with the help of spectrophotometer.

## EXPLAINATION

When the mixture is poured at the top of the column, the first layer of granules, which is surrounded by a small volume of liquid absorbs the solute. As the solvent front advances and the influx of new solvent (elute) from above dilutes the dissolved material to less than a certain concentration, the solute is eluted i.e. the absorbed molecules are, again released from the first layer into solution and carried over and absorbed on the next unoccupied layer of granules. This process is, further continued by fresh addition of elute and the solute molecule are removed down the column.

## THE ELUSION RATE

The rate of travel of the substances is determined by distribution co-efficient i.e the preference for either solvent or the adsorbent and is denoted by a numerical value. This value is more than 1 for fast travelling substances and less than 1 for slow moving ones.

For a particular set of adsorbent and solvent, the coefficient, even with a slightest change in structure of substance, get changed and this is only the reason, that is why, the adsorption process is so sensitive and useful for separating compounds.

As such the small variations in distribution co-efficient are translated into appreciable differences as the solvent travels down the column. Accordingly the different substances in the mixture move down the column at different rates & form separate zones.

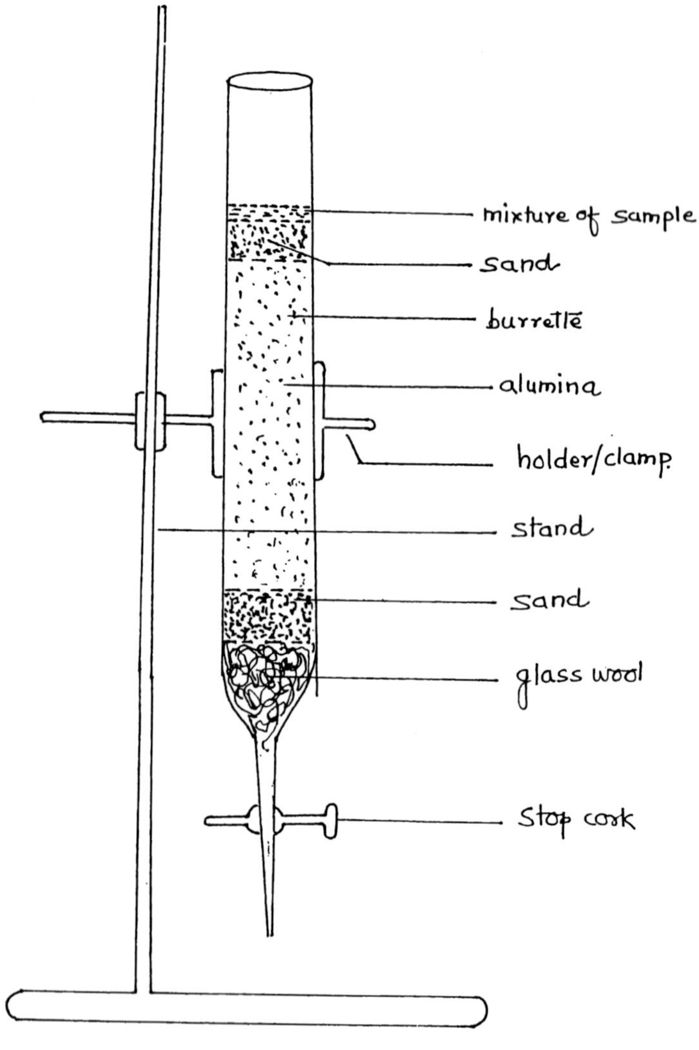


**Fig. 16.1 : Column Chromatography**

## Chapter -17

# PARTITION CHROMATOGRAPHY

### GENERAL INTRODUCTION

This is also known as ***liquid-liquid chromatography***. This technique is somewhat different from Tswett's solid liquid chromatography. It was discovered by British Scients Martin and Synge, a little after discovery of Tswetts'. In this technique, the column is packed with solid such as silica gel or cellulose, which is capable of absorbing water. The solid material like silica gel or cellulose is soacked in water before it is packed in the column. The substance (mixture of compound(s)) is dissolved in such a solvent like butyl alcohol, which in turn, runs through the column.

The different compound of the mixture get separated by virtue of their distribution coefficient between the ater held by the silicea gel and the solvent (water immisible) butyl alcohols flowing through the column. In other words, the different compounds are absorbed by water according to their distribution co-efficient, instead of being adsorbed by the adsorbent in Tswett's column.

This method is useful for separation of large number of water soluble compounds like amino acids, haemoglobins, pitultary harmones, carbohydrates etc. Separation of amino acids is, usually also done by this method. The separation depends on their distribution co-efficient between the solvent and water. Since cellulose or silica gel will take up appreciable amount of water, it is reasonable to propose that the amino acid dissolve in the water of the column and are extracted from it by the following solvent in proportion, to their relative affinities for the two phases.

## ION EXCHANGE CHROMATOGRAPHY

This technique is based on the mechanism of ion-exchange ability between the ion exchanger substances and the compound under treatment for separation.

The column is prepared of such exchangers like resins which may be either anion exchanger or cation exchanger. It is believed that the functional groups of the resin react directly with the various substances place on the column, but to a different degree. Thus, a resin containing acidic functional group, may be expected to react (directly with the various substances, placed on the column) with a mixture of bases placed on the column in the same way the bases will be distributed along the column, the most reactive base at the top and upon elution they will merge from the column in order of increasing basicity.

## PRINCIPLE

The polymerized resin (insoluble) commonly used for chromatographic purposes act as anion (-) or cation (+) exchanger. Because they have one of the ion (anion or cation) attached to the polymer and the other counter part in free to be attached to the compound under treatment. Accordingly the resin may be an anion-exchanger, if it has +ve ion fixed to the polymer and -ve ion is free or it may be cation exchanger it if has a -ve ion fixed and +ve ions is free.

The chromatographic capacities of the different resins, thus vary depending on the next ionic forces, they posses. Ion exchange resins are highly insoluble synthetic polymer containing functional groups, which are either acidic (-COOH, $-SO_3H$) or basic ($-NH_2$,-$N^+(CH_3)_3$ etc.) Those with acidic group-Acidic such as Dowex 50, Amberlite IR 100, are known as cation exchange resins. Those with basic functional group such Dowex 1, Amberlite IP4 are anion exchange resin. Some of the ion-exchange resins are given in the following Table 17.1.

**Table 17.1 : Ion exchange resins**

| *Trade name* | *Nature of Polymer* | *Nature of ion exchange* | *Potential strength (mg/ml.)* |
|---|---|---|---|
| Dowex 50 | Sulfonated polysterine | strong cation exchanger | 1.9 |
| Amberlite IR-12 | Sulfonated polysterine | strong cation exchanger | 1.9 |
| Amberlite IR-63 | Condensed acrylic acid | Weak cation exchanger | 4.2 |
| Dowex 1 | Polysterine with $CH_2N\text{-}Me_3Cl$ | Strong an-ion exchanger | 1.2 |
| Amberlite IRA 400 | polysterine with $CH_2NMe_3Cl$ | strong anion exc-hange | 1.2 |
| Dowex 3 | Polysterine with secon-dary amine | Weak anion exc-hange | 2.0 |
| Amberlite IR 45 | Polysterine with secon-dary amine | Weak anion exc-hange | 2.0 |

## MEHANISM OF ACTION

The action of resins is best understood by considering them to be polyfunctional acids or basis which, to be insoluble. Thus, if a solution containing an amine salt is placed on a column packed with a cation exchange resin in the solid form, the resin will take up the amine cation from the solution and release it with the hydrogen cation. The amine may then be subsequently displaced by washing the column with a more acid solution.

If a mixture of bases where initially applied, they could be individually released by the use of solvents of varying acidity and electrolytic content. Similarly, a cation exchange resin in the sodium form will replace such ions as calcium and magnesium in solution to form sodium salt. This is a principle, used in the softening of hard water.

The mechanism of ion exchange is possible due to cross lined structure of the resins. The water molecule and small ions easily penetrate the cross lines and the free ions are inter-changed e.g. if NaCl or KCl is treated with cation exchanger (fre +ve), the +ve ions of KCl/NaCl i.e. $Na^+$ or $K^+$, will be attached to the polymer of the resin and the +ve ion of the polymer will combine with $Cl^-$ to form HCl as shown in following equation:-

$$\text{Resin-RSO}_3^- \text{ —— } H^+ + Na^+Cl^+ \text{ —— } R.SO_3^- \; Na^+ + HCl$$

This exchange is due to the strong attraction of $Na^+$ or $K^+$ to the relatively charged resins. Similarly, the ion exchange resins attract the relatively charged ions of the compound as shown below :

$$R.\; NH_3^+ \cdot OH^- + Na^+Cl^- \text{ ———————— } R.\; C\; NH_3^+\; Cl^- + NaOH$$

## USES OF ION EXCHANGING RESINS

The use of ion exchange resins is not confined to chromatography, but to the other fields also like biochemistry and medicine e.g. resins are used for the prevention of blood clotting by decalcification of blood. To control gastric acidity and the dietary adsorption of electrolytes from the intestinal tract by the oral administration of suitable resins, the production of low sodium milk for special dietary purposes and the removal of $Sr^{90}$ from milk etc. It is also used in the field of wastewater treatment.

## AMINO ACIDS

The separation of amino acid can also be understand on the same basis. All the amino acids contain $COO^-$ group as well as $NH_2^-$ group, but they are either positively charged or negatively charged, depending on the pH. Accordingly they can be separated by the corresponding charged exchangers.

The cation exchanger separates the acidic, hydroxy or neutral amino acids by the use of acidic buffer (in which the amino acids are +vely charged), whereas the anion exchanger separates the basic amino acids by using alkaline medium, in which the amino acids are negatively charged the charged amino acids are then eluted with buffers of eluted with buffers of different pH in order to neutralise their charges and separate them from the bands of ion exchanger, the resin.

Experimental steps are as follows :

1. Select any of the earlier mentioned resins, depending on the nature of the amino acids(s), to be separated e.g., the analytical grade resin (Dower 50 X having 200-400 mashes) is selected for the separation of aspastic acid and proline.
2. Take distilled water in a container of a volume to fill the prepared column of 30 × 0.9 c.m. Suspend the resin (Dower 50 × 8 of 200-400 mesh size) and watch that the heavier particles are settle down.
3. Decant this water and leave the heavier particles behind and fill it into the vessel to form a column.
4. Prepare a citrate buffer pH 3.4 (0.1 M citric acid 40 ml + 0.1 M sodium citrate 10 ml). Allow nearly double volume of this buffer to run through column at the rate of 5ml/10-15 min.
5. Layer the sample mixture on the top of the column without disturbing the column (The sample mixture is prepared by shaking 2 mg each of aspartic acid and proline later the pH of this mixture is adjust to 2.5 by adding HCl).
6. When the sample has percolated through the column, run slowly distilled water in order to give sufficient washing to the column.
7. Hook the column to the function collector and the buffer receiver.
8. Elute the column with the citrate buffer and collect in tubes nearly 10-20 fractions of 5 ml. each.

**Assay of the fraction sample**

The fraction collected in the above manner are, then treated in the following way to assay them out -

1. Prepare a reagent of nin-hydrin by dissolving 2 gms of nin hydrin in a mixture of 25 ml. of methyl cellulose and 25ml. Na acetate buffer of pH 5.5. It can be stored in dark for further use.

2. Take 1 ml. of each fraction of separate tubes and add to each of them 1 ml. of nin hydrin reagent.
3. Heat the tubes in the boiling water bath for about 15 minutes and then wait till they cool down.
4. Add 5 ml. of diluting mixture (alcohol + water 1 : 1) and read the colour in calorimeter or spectrophotometer at 540µ. (green filter range) and 420 µ (blue filter range).
5. Note that the aspartic acid-nin hydrin complex absorbs at the 540 µ (green filter) and proline-nin hydrin complex absorbs at 420 µ (blue filter) wave length.
6. Make the fractions of the series, where blue and yellow complexes formed. In this way, two amino acids can be separated.

## AUTOMATIC AMINO ACID ANALYSER

This instrument is very perfect in the sense that it not only separates the amino acids quantitatively but also makes their quantitative estimation. All the amino acids (normally twenty) present in a protein can be separated by this instrument.

In this instrument, both types of ion-exchangers are used in a judicious combination. The cation exchangers separate the acid, hydroxyl and neutral amino acids by the use of acids buffers whereas the anion exchanger separates the basic amino acids.

## APPLICATION OF ION EXCHANGE CHROMATOGRAPHY

### Exchange of ionic constituents

Cation that constitute hardness in water (mainly calcium and magnesium ions are exchanged for an equivalent amount of sodium ion from a zeoline e.g. permulit water softer). Total salt content of a solution can be determined by passing the solution through the hydrogen from a strong acid (sulfonic type) resin, the displaced hydrogen ions from the resin is titrated by standard alkali, conversion of sulphate to chloride or of either of these to hydroxyl ions (or the reverse exchanger) can be made with anion exchange resins.

Chapter -18

# PAPER CHROMATOGRAPHY

## INTRODUCTION

This technique is widely used in the laboratories for quantitative isolation and identification of several compounds such as sugars, amino acids, nucleotides etc. The basic principle in this technique is very simple. The method was described for the first by Condsen, Gorden & Martin in 1944.

According to them, a small drop of solution containing the mixture of compounds, it is desired to separate is evaporated to dryness on a strip of filter paper and a suitable solvent system is allowed to flow slowly along the filter paper over the spot, either aided by gravity (descending) or by capillary alone (ascending).

The substances in the initial spot are extracted by the flowing solvent and carried along the filter paper to an extent which appears related to this distribution between the force and bound solvent phase of the filter paper, as discussed in relation to the column chromatography.

After the solvent system has flowed for a suitable distance along the paper, the paper is removed, freed of excess solvent by quickly drying and subjected to suitable test to locate the various compounds under study.

## AREAS OF APPLICATION

The technique of paper chromatography has became very popular by virtu of its simplicity, reproducibility and dependability.

In addition, very small quantity of test sample is required. It has found wide application in various fields of science engineering, medicine, agriculture & technology.

## TYPES OF PAPER CHROMATOGRAPHY

It can be done in any of the following ways-

a) ***Ascending or descending paper chromatography***

In this, the solvent is run on the filter paper either by gravitational phenomenon (in descending) or by the capillary action (in ascending).

b) ***Circular paper chromatography***

In this, the solvent moves horizontally from centre to periphery by diffusion phenomenon on a filter paper of circular shape.

c) ***Two dimensional paper chromatography***

In this technique, two chromatograms of the same mixture sample are obtained on the same piece of square filter paper. By running two different solvents, first in one direction and the other at right angle of the first. This technique is useful for separating large number of compounds in a sample mixture.

## PRINCIPLE OF PAPER CHROMATOGRAPHY

The basic principle is the same as that of partition column chromatography, but the phenomenon involved in paper chromatography are somewhat different.

Here, the solvent moves either by diffusion or gravitation/ capillary action on the filter paper.

The filter paper used in chromatography is made up of cellulose and is hydrated during use. Since most of the compounds have different solubilities in aqueous phase and organic phase, the property is responsible for separating them differently between two phases.

When the mixture of compound is spotted on the hydrated filter paper and the solvent is allowed to run through, the solvent, when moves over the spotted mixture, dissolves the variation compounds, according to their solubilities and hence moves at a different rate and occupy different positions in a given period of time.

The compounds, which have greater solubility in organic phase move at a faster rate on the hydrated filter paper in comparison to those which has greater solubility in aquous phase e.g. aliphatic amino acids like leucine or isoleucine have greater solubility in organic phase and move at faster rate as against the water soluble amino acids like glutamic acid or aspartic acid.

Thus, if a mixture of amino acids contains leucine & glutamic acid or aspartic acid, leucine will form a band ahead of glutamic or aspartic acid, when separated through this method.

## EXPERIMENTAL PROCEDURES

### I. Ascending or descending chromatography

Special air tight chromatographic chambers are available for this technique, but in the absence of such chambers, jars can be used for this purpose.

The following steps are to be followed-

1. Prepare any of the following solvents commonly used for the paper chromatography
   a) Alcohol, water and ammonium hydroxide in the ratio of 80:20:1 respectively.
   b) Butanol, acetic acid and water in the ratio of 4:1:1 respectively.
   c) Phenol and water in the ratio of 4:1 respectively.
2. Prepare the chamber for chromatographic use-
   a) For ascending technique, the prepared solvent mixture is poured into the chamber or into a large petridish kept within the chamber, to the extent of a few millimeter height.
   b) For descending technique, the solvent mixture is filled in a tube provided at the paper partition of the chamber. Some solvent is also put into the petridish. kept at the bottom of the chamber. The chamber is closed tightly so that it gets saturated with the vapour of the solvent kept inside.

3. Preparation of filter paper strip is the next step. Cut the strip of filter paper (Whatman No. 1) in accordance with the size of the chamber or jar. For ascending technique, the filter paper is taken in the square form, so as to form a single roll of it.

4. A pencil line is drawn on the filter paper 1-2" from the margin of the paper and the mixture of compound is spotted on this line at distance of 1" with the help of micropipette.

5. Now, open the chamber and fix this filter paper in the following way-

   a) ***For ascending technique***

      make a roll of filter paper and let it stand on solvent in such a way the spot remain slightly above the level of the solvent.

   b) ***For descending technique***

      dip the proximal end (spotted side) in the solvent mixture of the trough or tubular channel and hang the rest of the strip. In both the cases, care should be taken that the marked line remains just above solvent.

6. Close the chamber and allow the solvent to run through the filter paper to whole of its length.

7. Remove the paper with the help of forceps & hang it into air to let it dry. When the solvent is completely evaporated, put the paper into a hot air oven at about 80°C for a few minutes in order to dry it completely.

8. Now, spray the developer on the chromatogram.

9. Dry up the paper by handing it in air over night, if in hurry, dry it by keeping in the hot air oven at 100-110°C for 10 minutes or so.

10. The spots of individual amino acid & the mixture will be visible separately on the paper strip.

11. Measure the distance of the spots from the marked line and calculate the $R_f$ value for each.

$$Rf = \frac{\text{distance travelled by the substance}}{\text{Distance travelled by the solvent}}$$

i.e.

$$Rf = \frac{\text{distance of spot from the marked line}}{\text{distance of rear end of paper from line}}$$

## APPLICATION

The $R_f$ value of different amino acids are fixed under the same conditions and solvents. Thus the observed $R_f$ value in the experiment can be compared with the standard $R_f$ values and the conclusion can be drawn. The quantitative estimation of the (spots) substances can be done with the help of densitometer or some other analytical techniques.

## CIRCULAR PAPER CHROMATOGRAPHY

The principle is the same as for the ascending or descending paper chromatography, but the solvent run horizontally through the radii of the circular filter paper. This method can be performed without the chamber. The following steps are to be taken :

1. Take a large size petridish, clean it well and then fill it partly with the solvent mixture.
2. Take a circular piece of paper of the radius a little less than that of a petriadish & draw a circle of 2 c.m. radius in the centre.
3. Spot out samples along this circle with the help of micropipette.
4. Take a small piece of filter paper & make a thin cylindrical wick and insert it through a small whole in the centre of the circular filter paper, so that the wick forms a central stand.
5. Adjust the wick and the circular filter paper in the petridish in such a way that the wick is dipped into the solvent but the circular paper is held up horizontally above the surface of the solvent filled in the petridish.
6. After, sometime it will be observed that the solvent moves through the wick by capillary action & reaches the constant point of the circular paper. The solvent, then runs radially through the circular paper carrying the compound of the spots with it.
7. When the solvent has moved to the periphery of the circular paper, dry it & then develop the spots by spraying in the same way as done for ascending or descending paper chromatography.

## APPLICATION

Circular paper chromatography is ideal for the continue analysis of urine for estimation of sugar or separation of galcatose & glucose. This technique can also be employed for the screening of large number of amino acids distribution.

## III. TWO DIMENSIONAL PAPER CHROMATOGRAPHY

Sometimes, separation of some substances is not adequate when treated with any of the earlier mentioned methods. In such cases the chromatogram after having been developed in one direction is developed in the direction of 90 degree to the first direction, by using different solvents.

The following steps are taken in this technique-

a) Take a square sheet of filter paper of the size convenient for handling in chromatographic chamber.

b) Spot the mixture of compound in one corner of the sheet, 1-2" away from the both the edges of the paper.

c) Now, follow the steps as used in descending paper chromatography and develop the chromatogram but don't develop it and dry the paper.

d) Change the solvent (another solvent system).

e) Now fix the treated filter paper in the chamber in such as position that the solvent runs in the direction perpendicular to the first run (i.e. at $90^{0}$).

f) Develop and dry the paper.

g) Calculate the $R_f$ value of different spots and identify them with the help of standard $R_f$ values.

**Chapter -19**

# THIN LAYER CHROMATOGRAPHY (TLC)

## INTRODUCTION

This technique is employed to separate compounds by using thin layer of the separating element over glass plates. Now-a-days the technique of thin layer chromatography is universally used as a modern analytical method in chemical, biochemical, microbiological studies and research.

Stable (1958) was mainly responsible for bringing out the standard equipment for preparing thin layers, which helped this technique in becoming a modern tool of analytical chemical studies.

## AREAS OF APPLICATION

This technique is more advantageous, combining not only the advantages of paper & column chromatography, but in certain respect, it is superior to either of two.

Thin layer chromatography can be used both for organic and inorganic matters as well as on quantities ranging from the monograms to micrograms.

This technique has some specific advantages in the sense that it can be performed comparatively in a shorter period and by using only small quantity of the material with a result of greater sharpness of separation of hydrophillic and hydrophobic materials.

## PRINCIPLE

Thin layer chromatography is a technique, which involves all the three techniques simultaneously that a partition paper chromatography, adsorption paper chromatography and ion exchange paper chromatography.

The adsorbent material are quoted on a glass plate in the form of thin layer. These material are held together with a binding substance like plaster of Paris. These material are then developed with simple solvent system.

The use of inorganic adsorbent as layer material preferred in thin layer chromatography permits the identification of the fractioned substances even with the extremely aggressive sprays which due to the destruction of cellulose may not be applied in paper chromatography.

a further advantage is the low intrisic fluorescence of inorganic adsorbent which facilitates the identification of substances in ultraviolet light.

The lower detection limit of the analytical sample in thin layer chromatography is approximately one decimal power less than that in paper chromatography.

## EXPERIMENTAL PROCEDURES

Follows the steps given below :-

### a) Preparation of TLC Plates

1. Clean the glass plates thoroughly with water to remove the oil & greasy substances/material completely.

   Wipe item with alcohol and dry them in rack provided for the purposes.

2. Stretch the board or plastic support on the table and sprinkle a few drops of water on it.

   Then put the glass plates on the support in a linear fashion, the narrow plates are kept on either side.

### b. Preparation of coating material

1. For coating the plates, select any of the following adsorbent-

1. Silica gel G (with 200 meshes)
2. Alumina ($Al_2O_3$)
3. Kiesel glaker etc.

Silica gel G (with 200 meshes) contains 13% purified plaster of Paris & binder silica gel G F 254 contains a florescence indication upon exposure to ultraviolet light of 254 mv.

2. Weight 25 gms of absorbent & mix it with 50 ml. distilled water in a 150 ml. flask.

   Close the mouth of the flask with a plastic stopper & shake thoroughly for 30 seconds or more till a fine suspension is formed.

3. Transfer the absorbent suspension (slurry) from the flask into the applicator & turn the lever to $180^0$ to allow the slurry to ooze out.

4. Pull the applicators across the row of plates with a uniform motion (about 3 second per plate) without stopping until all the plates are covered with coating material.

5. When the applicator reaches the last plate, turn back the lever to stop the flow of slurry.

   If wrinkles appear in the thin layers, take each plate while still wet and gently bump to edge of the plate on a hard vertical surface.

6. Clean the applicator (spreader) immediately, brushing under tap water and rinsing with distilled water & then dry it in air.

**d. Drying of plates**

1. Leave the prepared glass plates on a level surface or on the template plastic support for superficial drying (about 15-30 minutes) until the surface of the layer becomes dull.

   Care should be taken that no dust is deposited on them.

2. Transfer these plates in a rack with guide (provided with the equipment) and dry them in an over for 1 hrs. at $110^0C$ for activation.

3. The activated plates are kept in dry place possibly in a desiccator, to protect them from deactivation.

**E. Application of the sample**

1. Special templates of plastic material are provided for the purpose of spotting the sample.

   The template has equally speed notches through which solution is applied to the plates with the help of micropiptetes or fine pointed rod.

2. Apply the sample in a row at starting line drawn gently on the plate.

   The size of the spot should be quite small (5μ), so that the solvent of the sample is quickly evaporated.

   To assist in this evaporation of the solvent, the plates may be warmed prior to the application or stream of air may be directly at the sample spot.

**F. Development of thin layers (plates)**

1. Prepare a solvent of n-Butanol, saturated with 0.05 percent sodium carbonate ($Na_2CO_3$) and place it in a trough into the developing chamber in order to saturate the chamber with the vapour of the solvent.

2. In most cases work is done in diffuse light. Sunlight should always be avoided.

   Compounds, that are extremely sensitive to light, should be applied in either red-or green light.

3. Allow the solvent to run through the thin layer of the plates

4. Dry the plates in air overnight or in hot air oven at 80°C for a few minutes in order to dry it completely.

5. Treat the plates with developer by spraying the solution over them with the help of sprayer.

6. Dray up the plates in air or in hot air oven till the spots become clearly visible.

7. Calculate the $R_f$ value of different spots.

## $R_f$ VALUE

### Retention value

The differences in the rate of movement of the compounds are caused by their various partition co-efficients, i.e. their different solubility in the mobile and stationary phases, the distance travelled by a substance in a certain solvent can easily be determined & when related to the distance travelled by solvent itself, is an intrisic property of the substance.

Mathematically :

$$Rf = \frac{\text{Distance of centre of the spot from the starting point}}{\text{distance of solvent front from the starting point}}$$

Since the course of solvent is always longer, the Rf value is always smaller than 1.

Sometimes, the $R_f$ value is represented by multiplying it with 100. Thus is denoted by the symbol PRf

Thus :-

PRf = Rf x 100

The $R_f$ value is a constant for each compound only under identical experimental conditions.

It depends upon a numebr of factors, of which, the following may be mentioned

a) Quantity of the layer material.

b) Thickness of the layer.

c) Activation grade of the layer.

d) Quality of the solvent.

e) Saturation of the chamber.

f) Presence of impurities.

g) Concentration of the substance applied,

h) Chromatographic technique (ascending, descending etc.).

Most of these factors can be eliminated by developing a chromatogram of a standard along with that of a substance. The distance travelled by a substance is compared with that of the standard. Therefore,

$$R_{st} = \frac{R_f \text{ of the substance}}{R_f \text{ of the standard}}$$

Unlike the $R_f$ value, the $R_{st}$ value may be greater than 1. This is because, when the substance is under investigation travels faster than the standard.

## Chapter -20

# GAS-LIQUID CHROMATOGRAPHY (GLC)

## INTRODUCTION

The gas liquid chromatography has more utility in studies of organic component at atmospheric particulates. With the recent advance in the gas chromatographic techniques like high pressure (performance) liquid chromatography etc. With the development of more and more sensitive and specific detectors, it has become possible to detect the organic compounds like polynuclear aromatic hydrocarbonds (PAH) and chlorinated hydrocarbonds like pesticides viz. DDT etc. to their lowest concentration of $10^{-12}$ gms (picogram) in the air.

Estimation of these compounds is important because many such compounds are highly toxic and carcinogenic. The mass spectroscopy coupled with gas chromatography may prove to be the most discriminating technique for the analysis of PAH, PCB and pesticides.

## PRINCIPLE

In chromatography, a mixture is separated by adsorption or solution partition between two immiscible phases. One is the moving while the other is stationary.

The moving or mobile phase may be either gas or liquid and the stationary phase can be either liquid or solid. Therefore, following chromatographic methods are possible :

1. **The *Gas Chromatography* (GC)**

   It is the general term used when moving phase is gas.

2. **The *Gas Liquid Chromatography* (GLC)**

   In which the mobile phase is gas and the stationery phase is an nonvolatile liquid which is coated into an inert solid support.

   The inert support and stationary phase together are termed as ***packing material.***

3. ***The Gas Solid Chromatography* (GSC)**

   in this stationary phase is an active solid material having surface active properties. Such solids material is called ***adsorbent.***

4. ***Liquid- Liquid Chromatography* (LLC)**

5. ***Liquid Solid Chromatography* (LSC)**

## SYSTEM'S CONFIGURATION

Graphic system consists of the following parts :

a. A column (or columns) packed with suitable packing material, kept in a thermostatic oven.

b. An injection facility with appropriate gas supply.

c. One or more detectors for quantitative analysis.

d. An amplifier and recorder for recording the signal from the detector.

e. Temperature programming facility to vary the oven temperature at desired rate. Column and detector switching facility to select required column and detector combination.

## PRINCIPLE

The sample injected into the column is carried by the carrier gas through the packing material.

In the process sample will remain partly in gas phase and partly dissolved by the liquid stationary phase. The ratio of concentration of sample component in the gas phase to that in the liquid phase is called ***partition co-efficient.***

$$K = \frac{\text{Concentration in the gas phase}}{\text{Concentration in Liquid Phase}}$$

Different components of the sample will have generally different affinities to the packing material, hence have different partition co-efficients.

Separating principle of the GLC depends upon the difference in partition co-efficients.

In GSC the separating principle depends on the variation in the extent to which the constituents of the sample are adsorbed on the solid adsorbents

Hence, GSC is also called ***Adsorption Gas Chromatography***.

## GAS SUPPLIES

The gas chromatograph required inert gas carrier to serve as mobile phase to carry the sample through the columns. It also requires suitable fuel and supporting gas for the detectors.

For FID and ECD, argon, nitrogen or argon methane mixture common used as carrier gases. The carrier gas used for TCD may be nitrogen or hydrogen. The carrier gas should be of ultra high purity. Generally carrier gas flow in the range of 20 to 100 ml./mt. is used.

Hydrogen is used as a fuel gas for the operation of flame ionisation detector, while compressed air or oxygen is used to support the flame. Supporting gas flow may be within 200-500 ml./mt., while hydrogen gas flow may be within 20-50 ml./mt.

## SAMPLE INTRODUCTION SYSTEM

Although there are other systems like bypass loops, automatic device etc., The most common and convenient system for introduction of sample is direct injection of a syringe (Hamilton type).

Sample volume in the range of 0.1 μl to 10 μl is generally injected. It requires some practice to get repeatable results.

## COLUMNS AND PACKING MATERIALS

Generally glass or stainless steel tubings of 1/8" (3.2 mm) to

1/4" (6.4 mm) capillary columns are still fine I.D. are used as GC columns.

Actual column length may be from 1 meter to 5 meters, but in order to house them in the oven, they are coiled in suitable size.

In some special cases, capillary columns of 1/16" or less are used. More the length of the column, better is the separation and more is the retention time, hence compromise is usually done while selecting the column length.

## ADSORBENTS FOR GSC

Most common adsorbents which are used as stationary phase in GSC are activated charcoal, chromosorb, activated alumina, silica gel, porapak (Q, N) and molecular sieves.

These are mostly used for separation fo atmospheric gases. Porapak-Q is suitable for NO, $CO_2$, $SO_2$ etc. Porapak-N is suitable for separation of acetylene from ethylene and ethane, while molecular sieve is suitable for $H_2$, $N_2$, $O_2$, CO, $CH_4$ etc. and silica gel is suitable for gases and light hydrocarbonds.

## SOLID SUPPORTS FOR GLC

In gas liquid chromatograph, the inert diatomaceous earths are used as solid support over the stationary phase-liquid supports are coated. Most common solid supports are chromosrob, P, G, W, gas chrome Q and Porapak resins.

## LIQUID PHASE OR LIQUID SUPPORT

Most common liquids which are coated on solid supports to serve as stationary phase as squalene (useful for aliphatic hydrocarbons, methyl silicon gum (SE-30), fluoro silicon (QF-1), Silicon Oil (DC-200), methyl silicon (DV-101), Apiezon greases (suitable for n-alkanes and halogenated hydrocarbons) and polythylene glycol (carbons).

## MESH SIZE AND LOADING

Depending upon the diameter and length of the column, mesh size of absorbents and solid support as well as loading of liquid phases may be decided. For 1/4" column 60-80 mesh absorbents (or solid supports) with 15% by wt. liquid support coating concentration is suitable, while for 1/8" column, 60-80 mesh and absorbents or 60-80 mesh solid supports with 10% liquid support loading is recommended. Higher mesh size is also often used to get more sensitivity.

## TEMPERATURE FOR OPTIMUM OPERATION AND REGENERATION

All the solid adsorbents, solid supports and liquid supports are to be operated below/within a maximum temperature as prescribed by their manufacturer. Hence, it is different for different material. The common operating temperatures are within the range of 50° to 250°C.

## OVEN AND TEMPERATURE CONTROLLING AND PROGRAMMING ARRANGEMENT

A well insulated stainless stell oven of suitable size to house two columns is used. Temperature of the oven is precisely controlled in the complete operating temperature range of columns 1.8, 50°–300° with an accuracy of ± 0.1°C.

## DETECTORS

The three most utilized detectors are thermal conductivity detector (TCD), the flame ionisation detector (FID) and the electron capture detector (ECD). The flame photometric detector (FPD) and alkali flame ionisation detector (AFID) are also often used in some special studies.

## THERMAL CONDUCTIVITY DETECTOR (TCD)

In thermal conductivity detector, platinum (or of other inert metal) filaments connected in a wheat-stone bridge is kept hot by passing through it a current of order of 100 to 250 amps..

When the different gaseous components of the sample are passed over the filament along with the carrier, because of the difference in their thermal conductivities, the temperature and hence the resistance of the filament varies and gives rise to change in bridge current which is amplified and recorded.

The TCD is most suitable detector for separation of gases. Out of all gases, $H_2$ is most conducting while argon is the least.

Although sensitivity of TCD is limited to only $10^{-7}$ gms, it does not require any fuel and supporting gases but only carrier gas is required which is an added advantage over other detectors with FID.

As TCD works on principles of difference in thermal conductivity of carrier gas and sample vapour, any slight change in operating temperature or carrier-gas-flow will affect its performance.

## FLAME IONISATION DETECTOR (FID)

This is most widely used form of the detector. It consists of a small stainless steel burner and then electro meter arrangement. The flame of the burner is maintained by hydrogen and compressed air.

Principle of the detector is that when different hydrocarbon components of the sample are burnt in this hydrocarbon components of the sample are burn in this hydrogen-rich flame, the extent of ionisation is different with different component depending upon its carbon ions.

The ionisation current is amplified and recorded FID is most for separation by hydrocarbons (Heptocyclic and Aliphatic), Polynuclear aromatic hydrocarbons.

Unlike the TCD, temperature regulation is not so critical. it requires additional gasses $H_2$ and air for its operation.

## ELECTRON CAPTURE DETECTOR :

In this detector, a thin cylindrical foil of β radio activity source like $Sr^{90}$ or $Ni^{63}$ is used. The β particles ionises the carrier gas (argon or $N_2$). When any electrophilic compounds from the sample enter the detector, recombination of electrodes or capture occurs and reduce the ionisaion current and the signal.

## CHROMATOGRAM

The time from the injection of sample to the elution (peak) of a particulate component is called the ***retention time***.

The retention time gives only quantitative information, because under the experience mental conditions, chromatorgaph depends upon the nature of the component.

## DUAL COLUMN TECHNIQUE

Very often two identical columns and two detectors are used in the differential mode detectors are used in the differential mode. This reducing the noise created because of the fluctuations in the set temperature and flow rates.

## STANDARDIZATION

This should be done using chromatographic pure chemicals and standards of the solvents and components to be studies. A standard chromatogram is obtained using optimum conditions of gas flows column and over column and oven temperature and the detector.

## SAMPLE PROCESSING

Because of the very low concentration of organic particulates matter for collecting the sample in air, high column sampler is used. Then it is extracted using organic solvents like benzen or cyclohexane, taking sufficient care that highly volatile low molecular weight part of PAH is not lost.

## Chapter -21

# GAS CHROMATOGRAPHY (GC)

## INTRODUCTION

### Difference between gas-solid and gas-liquid chromatography

The gas-solid chromatography employs a solid adsorbent as a stationary phase, whereas gas-liquid chromatography accomplishes a separation by partitioning a samples between a mobile gas phase and a thin layer of non volatile liquid held on a solid support.

### The sequence of a gas chromatographic separation

1. A sample containing the solute is injected into a heating block where it is immediately colourised and swept as a plug of vapour by their carries gas stream into column inlet.
2. The solutes are adjusted at the head of the column by the stationary phase the disrobed by fresh carrier gas. This partitioning process occurs repeatedly as the sample is moved the outlet by the carrier gas.
3. Each solute will travel at its own rate through the column and subsequently a band, corresponding to each solute will form. The bonds will separate to a degree, that is determined by the partition ratio of the solute and the agent of band spreading.
4. The solutes are eluted, one after another, in the increasing order of their partition ratio and enter a detector attached to the column exit.
5. If a recorder is used the signals appears on the chart as a plot of the times verses the composition of the carrier gas stream.

6. The time of emergence of a peak identify the component and the peak area reveals the concentration of the component in the mixture.

The gas chromatographic method is commonly use for volatile materials.

## COMPONENTS OF A GAS CHROMATOGRAPHY

The following units are present in a gas chromatography

1. a supply unit of carrier gas in a high pressure cylinder with pressure regulator and flow meters.
2. A sample injection system.
3. The separation column.
4. The detector.
5. An electrometer & strip chart recorder and,
6. Separate thermostat components for housing the column & the detector so as to regulate their temperature.

Helium is the preferred gas, but is not easily available & thus nitrogen can serve as an alternative.

## SAMPLE INJECTION SYSTEM

This is the most care taking part of the gas-chromatography. The sample must be introduced as a vapour in the smallest possible volume and in a minimum amount of time without either decomposing or fractioning the sample or up setting the equilibrium conditioner of the condition or both. Quantity of sample introduced & the nature of introduction must be reproducible with a high degree of precision (Fig. 21.1).

Liquid samples are usually injected by a microsyringe through a self-sealing rubber septum into a metal block that is heated by controlled resistance heater. Here the sample is vaporised as a plug and carried into the column by the carrier gas-stream. Because the entire sample must undergo instantaneous vaporisation to attain plug flow, the injection zone temperature must exceed the boiling point of all components.

Gas samples, although can be injected by a gas tight syringe. Loops are also available in volume, ranging from several microlitres to several millilitres.

A T-shaped injection port permitted syringe injection of a liquid sample into a sealed hot zone. After a short period of time (to allow for complete vaporization), the values are switched to sweep the sample from the hot zone into the column.

Well coated open tubular column can accommodate only thousands of a microlitre of sample. To obtain the desired sample size, a sample splitting device is placed between the injection block and coumn. A typical system uses a 0.5 $\mu l$ (microlitre) injection split 100 to 1, where 1 part enters the column & 00 parts are exhausted into the atmosphere.

Pyrolysis has become a widely accepted method for solids. In the tube type pyrolyser, the sample is placed in a sample-boat that is positioned in the centre of the tube. The tube is, then sealed and pulse heated electrically at a controlled rate of temperature rise to the desired final temperature. Then the tube is opened & the vapous swept into the column with the barrier gas stream.

Operation based on the Curie's principle utilise filaments of certain alloys, which when subjected to intense of energy, rapidly heat to a specific temperature, unique to the alloy. Rise time to temperature as high as 800°C are on the order of nano-seconds.

A constant flow of carrier gas is maintained in the sample. Because the filaments must be coated with the sample, this technique is restricted to material, that will coat out of solution.

## CHROMATOGRAPHIC COLUMN

The basic types of chromatographic column are, in general use-

a) Packed and

b) open tabular column

Packed columns are tubes that have been filled with an inert support coated with a non-volatile liquid phase for use in gas liquid chromatography.

Open tubular column differ from the packed column in that the gas path through the column is an unrestricted hole through the tubing and the separating medium is coated on the wall of the tubing. The pressure drop is orders of magnitude smaller than that of packed columns of the same length, this permits the use of very long column.

Packed columns normally are used in lengths of 0.7 to 2.0 mete; open tubular column run anywhere from 30 to 300 meter in length. Column are made of tubing coiled into an open spiral or a flat pancake shape.

The separating ability per minute of an open tubular column does not differ greatly from that of packed coumn, however, the use of very long column coupled what relatively rapid analysis times provides the chemist with a means of repeating compound that have small differences in their physical characteristics.

The major drawback to wall coated open tubular column is the small amount of liquid phase that the wall is capable of holding. The objection is overcome by increasing the surface area of the column by coating the wall with a finely divided support upon which a much larger amount of liquid phase can be coated, yet, without increasing film thickness. This is the support coated open tubular (SCOT) column.

The main advantage of an open tube, namely lower pressure drop that makes long lengths feasible, is preserved, but the β value, which is

$$\beta = \frac{\text{Vgas}}{\text{V sample}}$$

approaches that a packed column and column capacity is greatly increased.

The tubing is wider (0.5 mm), the flow rate faster (4-70 ml./min and the total volume of the connection is less critical sample splitting is useful but not required if sample size is 0.5 μ l or less.

If the separation demands more than 10000 plates a scan column is probably the best solution.

## SUPPORTS

The support in general use for packed column is diatomaceous earth which has been crushed & calcinised at 900°C (pink support) or which has been mixed with a small amount of flux $CaCO_3$ and calcined above 900°C (white support).

The best particles size is 100 much for 2 mm (1/8 inch) column and a 80/100 much for a 4 mm column for effective packing the inner diameter of tubing should be atleast 8 times of the diameter of support particles.

The white support has pore size of about 9 μ m, whereas the pink support has a pore size of about 2 μ m. A column of the pink support will held more of the liquid phase and hold it in smaller and shallower pools, which require shorter transit time for the solute.

Both types of supports perform quite well for the analysis of relatively non polar samples but the tend to be too active for the polar samples. The surface of diatomaceouis supports are covered with silanol (SiOH) group, that tend to absorb sample molecule particularly when lightly loaded with liquid phase or when non polar liquid phases are used. This effect can be reduced greatly, if the sinarol group are converted to silyl ether by reaction with dimethyl dichlorosilane. As a result, surface activity and peak tailing are reduced considerably. At the same time, however, silanization reduces the surface area of the support & limits stationary phase loading to the maximum of seventy percent. Thus, there is a constant compromise in the choice of solid support.

Solid support having surface energy low enough to present trailing are wetted poorly by the liquid phase, resulting in low efficiency. Those that are easily wetted usually cause tailing of polar samples.

Special supports are also available for certain applications. Very lightly loaded glass beads are used for very rapid analysis well below the boiling point of the sample components. The surface of the beads is roughened by several techniques to obtain better wetting and increased liquid phase capacity. Most phase loading are 0.05 to 0.20 percent of the liquid phase. It can be used for analysis of corrosive samples.

Porous polymers of beads are also used in gas solid chromatography. These beads have very low affinity for water and polar molecules, which, therefore, elute very rapidly from the column as symmetrical peaks. These polar polymer beads are sold under trade names, Poropak and Chromosorbs.

## LIQUID PHASES

Indeed, thousands of partition liquids have been described in the literature. It is true that many offer unique separations, it is also true that most separation can be performed on a relatively few liquid phases. Several of the more common liquid, partitioning liquids, ranging from non-polar to polar & their properties may be different.

Polar liquid phases are the best for polar samples e.g. Squalene for hydrocarbon, carbowax for alcohols & polyesters for fatty acids, methylesters. Liquid phases which are similar to the compound related to these components, compare to liquid phase, which are dissimilar. This aromatic compounds (induced polarity) are retarded on Carbowax, whereas paraffins (non-polar) are eluted more rapidly. Conversely, alcohols are eluted more rapidly on Sequalane relative to paraffins. This property can be used to make group separation or to accelerate or related components with respect to others in the sample.

Unless the sample dissolves well in the liquid phase, little or no separation occurs. The gas phase is inert & separation occurs only in the liquid phase.

A reasonable guidelines is that a retention of five time the air peak is required to show any reasonable selectivity. Lower temperature will almost always increase the solubility and selectivity.

The minimum temperature limit is important & is determined by the melting point or the velocity of liquid phase. As velocity increases & the mass transfer between the gas & liquid phases becomes so slow that little separation occurs.

The upper temperature limit of liquid phase stability and velocity has been a barrier.

## DETECTOR

Located at the exit of the separation column, the detector senses the arrival of the separated component as they leave the column and provides a corresponding electrical signals. The temperature of the detector component must be sufficiently high to the present condensation of the sample vapours, yet not cause sample decomposition.

### 1. Thermal conductivity cell detector (TCD)

The termal conductivity detector is made of four filaments arranged in an electrical bridge net work. Each helical filament is situated in a separate capacity in a brass block, which serve as a heat sink. The filament in opposite arms of the bridge are surrounded by the carrier gas, the other two filaments are surrounded by the effluent from the column. The temperate of the filament is determined by the rate of heat loss by conduction through the carrier gas. The gas flow around the filaments and through the cavities. As components elute from the column, the composition of the gas changes. The resultant change in thermal conductivity produces a change in filament temperature and, in turn, a change in the resistance of the filament, causing an electrical output from the wheat stone bridge circuit. Aside from hydrogen, helium has the highest thermal conductivity of all gases & as a carrier gas, provides the highest sensitivity of detection for all other gases.

Because of its simplicity, the thermal conductivity detector is preferred for survey work & for moderate sensitivity in all areas. It responds to all types of inorganic and organic compounds. Because it is non-distructive, it is particularly suitable for preparative or fraction collecting work. The cavity cell volume is 2.5 ml. for detector associated with 4 mm column, this is decreased to 0.25 ml. in the microcell designed for use with SCOT & small diameter packed column. One annoying initiation of the detector is its low resistance to oxidation and sometimes to chemical attack of the gold-sheated tungsten filaments. The detection limit is about 5 microgram per millilitre of sample gas concentration or 10 nanogram of sample weight & the range is about $10^6$.

### 2. Flame ionization detector (FID)

The flame ionization detector currently is the most popular detector because of its high sensitivity, wide range and great readability. It consists of a small hydrogen flame burning in an excess of air and surrounded by an electrostatic field.

Column effluent enter the burner (flame) base through a millipore filter and is mixed with the hydrogen entering the burner. Organic compound eluted from the column are burned. During the combustion ionic fragments and free electrons are formed. There are collected, producing an electric current proportional to the rate at which the sample enters the flame.

Many variation in design are available. Some instruments add make up gas velocity and sweep the column effluent through the 'dead' column between the column exit and the detector entrance, thus minimising extra column band spreading. The FID respond only to oxidisable carbon atoms and response is proportional to the number of carbon atom in the sample components. There is a no response from fully oxidised carbon such as carbonyl or carboxyl groups and their analogs and response diminishes within creasing substitution of halogens, amines and hydroxyl groups.

The FID does not respond to inorganic compounds apart from those easily ionised. Insensitivity to water. The permanent gas like carbon mono oxide (CO), carbon di oxide ($CO_2$) is advantageous in analysis of aqueous extract and in air pollution studies. Ofcourse, carbon mono-and di-oxides can be converted easily to methane by reduction with hydrogen over a nickel catalyst.

Detection is about 20 picrogrma of sample weight or about 5 nano-gram per millilitre of sample gas concentration. Linear range is $10^7$. Precise temperature control is not a requirement for this detector an obvious advantage in programmed-temperature applications.

**3. Electron-Capture Detector (ECD)**

The principle of the ECD is based on electron absorption by compound housing an affinity for free electrons. These are compounds that have an electronegative element or group upon exposure to a source of flow energy electrons. There compound tend to attach or capture electrons to form negative ions.

As nitrogen carrier gas flow through the detector β particles from the tritium source (or Nickle-63).ionise the nitrogen molecule & form slow electrons which migrate to the anode under a fixed potential which can be varied from 2 to 100 V. These collected elements produced a steady base line current flowing across the cell from A to B.

The production of secondary electrons only in a particular area. Methane is mixed with carrier gas and serve as a quench gas to reduce the energy of the element for more efficient capturing. When an electron-capturing component emerges from the column and moves from C to E, it react with an electron to form either a negative molecular ions or neutral radical and a negative ion. These are swept out by the gas flow. The net result is the removal of an electron from the system and a decrease in the standing current. The decrease is recorded as a negative peak on the recorder.

A pulse power supply with a pulse duration only long enough to collect electrons (but ignore the negative ions) is commonly used.

Response is essentially non linear flowing an exponential relation similar to Beer's law. By using feedback and varying the pulse rate, operation is linearised, giving a linear range of 500 detection is at the 0.1 picogram level.

More important is the selectivity towards traces of halogen containing compounds, anhydrides, peroxides, conjugated carbonyls, nitrites, nitrates, ozone, oxygen organomethyls and sulphur containing compound whereas the detector is virtually incentive to hydrocarbon, amines & ketones.

Selectively ratio of 100000 to 1 are quite common. The most common application fo the ECD is the determination of chlorinated particles. Others are halogenated anaesthetics polynuclear carcinogen and sulphur hexafluoride traces in metereology.

Hydroxy or amino compounds can be converted to hepta flurobutyryl or tetrafloroacetyl derivatives which are helpful in Biomedical works.

### 4. Other detectors

***The flame photometric detector***

Employs a photomultiplier tube to view the emission from a reversed, fuel rich hydrogen flame through an interference filter.

Single detectors are used either for the Hg band system of P at 5260 Å or the $S_2$ band system of S at 3940 Å.

Alternatively a dual detector monitors both signals simultaneously. The burner assembly is similar to that used in the flame ionisation detect except that air and carrier gas flow through the jet and hydrogen is supplied by a separate annulus around the jet.

Response to P is linear over a range of about 10000 to 1 with a detection limit of 10 picogram. S response is proportional to the square of the quantity and is nearly linear on a log-scale over the range from 200 picogram to 100 nanogram. This detector finds use in air pollution studies for sulphur entities and pesticides analysis for phosphorous.

A regulator sprayer burner or slot burner plus appropriate interference filters or gratings monochromotors & photommultiplier detector, can be used to detect many other organic elements, including silicon from silanized derivative.

Other detectors are also available but not used as widely in gas chromatography. The gas chromatography has been combined with other instruments, particularly the Mass-spectrophotometer.

## DUAL DETECTORS

In dual detector, operation of the effluent from a single column is split and is sent to two different detectors simultaneously. A two-pen readers allow simultaneously presentation of the signals. Two detectors, such with their individual selectivity, make possible a more detailed characterisation of the sample than can any single detector.

It aids in quantitative analysis by distinguishing components comprising overlooping peaks.

In the differential mode, matched columns and identical detectors are used. This arrangement eliminates baseline drift caused by bleeding of the liquid phase, particularly during temperature programming.

## QUANTITATIVE MEASUREMENT

Ordinarily the chromatogram is traced by a strip chart recorder connected to the output of the detector amplifier. There should be three condition must be provided-

a) The output of the detector-recorder system must be linear with concentration. The range is expressed by the usable range of the detector and coupled with sensitivity, gives the concentration limits.

b) The carrier gas flow rate must be constant so that the time abscission may be converted to volume of the carrier gas.

c) The pen response of the recorder must match the speed of the detector's response or one must couple automatic integration devices directly to the electronics of the detectors. In other words, if the peak width is 0.2 second one can not use a recorder with a second response time. Under these conditions, peak area can be used as a quantitative measure of component present.

## CORRELATION OF AREA AND QUANTITY

Area under a gas chromatographic peak can be correlated. Similar areas obtained with known concentration of the particular solute run under condition that are as similar as possible this is usually done by the construction of a graph of quantity versus peak area.

Errors in the injection technique normally limit the accuracy of 0.5 percent at best. To express the result of the analysis in percentage, one can assume that all of the components of the sample are present in the chromatogram.

The sample size can then be obtained by summing up the individual components. Individual quantities are referred to this sum. This procedure is termed "normalization"

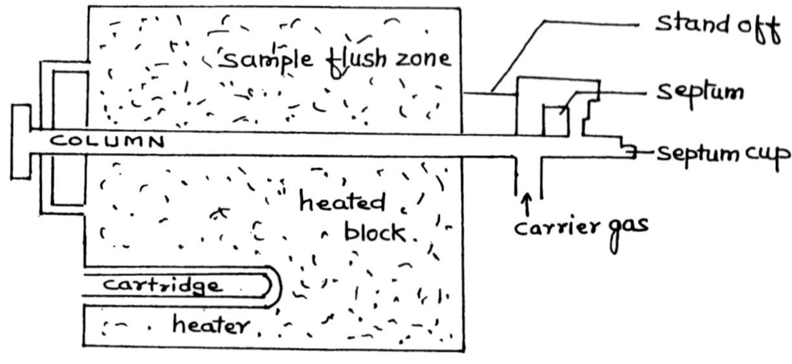


**Fig. 21.1 : Cross section of sample inlet system**

Chapter -22

# HIGH PERFORMANCE LIQUID CHROMATOGRAPHY (HPLC)

## INTRODUCTION

In 1906, Tswett, a Russian botanist separated components of plant pigment on calcium carbonate column using petrol ether as mobile phase to invent a new technique of separation to which he called ***chromatography or colour writing***.

However, inspite of lot of work done by Tiselius, Martin, Synge etc. in liquid chromatography, gas chromatography became the primary technique of microlevel analysis for volatile organic compounds.

With the cross fertilization of ideas between various chromatographic techniques, around late sixties, liquid chromatography reappeared in the form of high performance liquid chromatography (HPLC). Unlike gas chromatography, HPLC can be used to analyse non volatile compounds, highly polar compounds, ionic samples and even heat sensitive biological molecules. In liquid chromatography components of a mixture are separated due to their differential distribution between stationary and mobile phase.

In gas chromatography mobile phase is generally an inert gas and functions only as a carrier of vapourised sample but in liquid chromatography, mobile phase participates actively in the sample components which offers an additional variable to control the separation and expand the range of analysis. Because of higher viscosity of liquid the operating pressure required of HPLC (3000 psi) are about hundred times more than in case of gas chromatography. Lower rate of diffusion requires particle size of the packing (≈5μ) about twenty time smaller. Both these factors caused considerable delay in the development of HPLC.

## CLASSIFICATION

Liquid chromatography can be classified depending upon the forms of the stationary phases into thin layer chromatography, paper chromatography, column chromatography etc. Modern liquid chromatography (HPLC) is faster, cheaper, simpler and much higher version of the classical chromatography.

HPLC employs various types of interaction such as chemical (Ion-exchange), physical (adsorption, partition), mechanical (size exclusion) for the separation of the sample components.

In adsorption chromatography, the stationary phase is an adsorbent and the separation of is affected by the repeated adsorption desorption (generally silica gel or Alumina) and a mobile phase (low polarity solvent such as hexane, chloroform etc.)

In partition chromatography, the separation is achieved by partition of the components between stationary, liquid phase coated or bonded to stationary packing and mobile liquid phase. When the stationary phase employed is polar with amine or cyano groups attached to the surface of the packing and mobile phase is less polar such as hexane, isopropanol etc. then is called ***normal phase chromatography***.

In this case, polar compounds are retained on the column and hence elute later to non-polar components. In reversed phase chromatography (RPC) stationary phase is non polar silica with octadery saline or other similar functions bonded to the surface and mobile solvent is highly polar like water, methanol or acetonitrile.

In this mode, non polar compounds are retained longer vis-a-vis polar ones. Since most of times, active functional groups are chemically bonded to the inert packings, they are also referred to as ***bonded phase chromatography (BPC)***.

In ion-exchange chromatography, the stationary phase has an ionically charged surface and mobile phase is aqueous buffer with controlled pH and ionic strength. Ionic species are separated by competition of sample ions with counter ions in mobile phase for the charged sites on stationary phase.

Stronger ions stay longer in the column, whereas weaker ions elute early. The technique can be used for the analysis of cations and anions. In ion-pair chromatography, the sample ions are modified to form neutral ion-pairs either in the mobile phase or on the pecking surface using pairing-reagents. The neutral pairs are then separated by partition chromatography.

In size exclusion chromatography, the column is filled with silica or polymer gets with controlled pore sizes and the sample is simply screened or filtered according to molecular size differences as it is washed through the column. Larger molecules effecting first as they can not be retained in pores.

## PRINCIPLE

When the inert sample/solute is introduced into the mobile phase stream at the top of the column, the solute molecules would travel through the column with the same speed as the mobile phase.

The time taken by it to travel through the column is called ***mobile phase***. However, if the solute molecules interact with the stationary phase, time taken by it to travel the column is called ***retention time***. During its passage solute molecules will spend some time in the mobile phase (atm) and rest of the time on the stationary phase.

Thus adjusted retention time $t_R{}^1 = t_R = t_m$

capacity ratio k is defined as the ratio of the time spent by the solute on the stationary phase to the time spent in the mobile phase:

$$K = \frac{T^1R}{t_m} = \frac{T^1R - t_m}{t_m}$$

In practice, capacity ratio of the first peak of interest should be about 1, for it to be separated from the solvent and its impurities and not to have peak for K more than 12 as peaks become too broad for good resolution and sensitive detection.

As the solute travels down the column, because of longitudital diffusion its band broadens and the peak obtained has the form of Gaussion Curve, the peak width indicates the quality of column, its value depending upon the flow rate among other parameters.

Therefore, the column efficiency is expressed by a parameter, number of theoretical plates or plate number n as :

$$n = 5.545 \left(\frac{t_R}{W\lambda}\right)^2$$

$W\lambda$ being peak width at half height.

Plate height equivalent to theoretical plate, $h_l$ considers the length of the column L as well :

$$\lambda = \frac{L}{n}$$

These terms originate from the theory of distillation, the plates here not being real. Lower plate height or higher plate number indicating higher efficiency column. In practice, generally one encounters 25 cm. long column with plate number around 5000.

Separation factor of two solutes ended in the sample ($\alpha$) describes the relative positions or two peaks as -

$$\alpha\ \beta A = \frac{Kg}{KA} \text{ or } \frac{K_{RB}}{t_{RA}}$$

$\beta$ is being later eluting solute when $\alpha = 1$ the components A and B are not separated. Separation factors depends upon the stationary phase, mobile phase, temperature etc. Peak resolution depends upon the separation factor indicating the position of the peaks and on the sharpness of the peak determined by the efficiency. Conventionally, peak resolution R is defined as :

$$P = \frac{2\partial t}{W_a+W_b} \text{ or } \frac{t}{W_B}$$

t being the distance between two rating and $W_A$, $W_B$ their full widths at half height. The value of R = 2.5 represents baseline separation whereas R = 1.7 indicate about 90 resolution.

Velocity of the solute molecule through the column $\mu$ = L/tm greatly affected the analysis, if too low, solute molecules will spend too much time in the mobile phase and if it is too large, solute molecules may not have enough time to interact will all active sites on the stationary phase (Fig. 22.1).

The plot of velocity $\mu$ versus the plate height called van deem ter curve as given in Fig. 22.2. Best column efficiency, practically $\lambda$ = 0.015 can be obtained at the minimum of the curve cc responding to solvent velocity of about 0.5 nm/sec.

However, usual working velocities are around 2-5 nm/sec. corresponding to 0.5 to 1 ml./minute. flow through 0.6 nm to column. Because the controlled porosity and small particle size of the packing even at these velocities loss in the column efficiency is less.

## INSTRUMENTATION

Fig. 22.3 presents the functional schematics of an HPLC system. Basi components of such a system are mobile phase reservoirs, a pump to proper the mobile phase, an injector for sample introduction, a column containing the stationary phase, a detector to determine the components, a recorder or data processor to collect and display the data and a fraction collector to collect the separated components.

For gradient elusion and flow programming one more pump and controlling unit is required whereas column oven also is employed for some applications. Mobile phase reservoirs either of glass or inert polymer store. Various high purity solvents without contamination.

Since the solid particulates can block or damage the system components, solvents are prefiltered by membrane filters. Arrangement is made to remove the dissolved gases either by sparging, sonification, evacuation or boiling.

In HPLC systems reciprocating positive displacement pumps are generally employed. Here a rotating eccentric-cum-forces the piston to expel the solvent through a one way check valve, pumping rate is controlled by adjusting the length of the stroke. Pulse dampers in the form of the Bourdon pressure gauge is used to smooth the flows.

Dual head reciprocating pump gives much better flow. To programme the mobile phase composition two or more solvents are blended either on the low pressure side or high pressure side of the pump. In high pressure mixing, system output of two independent pumps are connected to a 'T' shaped mixing connector and the pumping speed is compressed by solenoid operated proportioning valves.

The solvent composition can be varied linearly or non linearly (convex or concave curves) and the programme parameters can be adjusted by the solvent programmer. Modern programmer can handle even multi level, multi-component elution programmes. Presently, commercially available pumps can deliver accurate solvent flows in the range of 0.01 ml./min. to 10ml./min. upto 5000 psi. They have low internal volumes and are inert in most solvents.

Sample is introduced into the system by low volume septless injector valve which can stand pressure upto 6000 psi. when the valve is in bypass position sample is introduced into 100 ml. sample loop by the syringe and solvent is flowing through the column.

In 'inject' position concern of the loop is flushed into the column. The separation column are generally made up of 816 stainless steel. Standard columns are generally 25 cms. long with 1/4 inch O.D. (outer diameter) and 2.6 to 3 mm or 4.6 to 5 mm I.D. (internal diameter).

Swageok$^R$, Gyrolok$^R$ compression fitting and 1/16" as tubings are generally employed for the connections. Most widely used packing for liquid partition chromatography have stationary phase chemically bonded to support particles by chemical reaction between the surface hydroxyl groups of silica particles and a linear organic molecules or an organosilane.

Polar bonded phases have amino or cyano groups of the end of the hydrocarbon chain whereas non polar bonded phase have alkyl chain (e.g. octadecyl) bonded enough the silicon atom of alkylsiline.

Two types of support particles are employed. First type is the porous through to the use other system. Second, pellicular has solid core with a thin porous skin.

Totally porous support particles generally consist of silica get with high surface area and are available in wide variety of particle sizes (3-50 μ) whereas pellicular support consists of glass beads coated with the thin skin of silica, regular in shape, but available in relatively larger sizes only (30-50 μ).

Presently three categories of columns are used frequently :

1. Standards analytical column
2. Microbore high resolution column and
3. Fast LC column.

Comparatively all three of them are presented in Table 22.1.

**Table 22.1 : Performance comparison of the columns**

| *Specification* | *Standard* | *High Speed* | *Microbore* |
|---|---|---|---|
| Length (mm) | 100 | 60 | 100 |
| Internal diameter (mm) | 1.6 | 4.6 | 2.1 |
| Particle size of the plate (μ) | 5 | 3 | 5 |
| rate (ml/min.) | 6000 | 6000 | 6000 |
| Volume (ul) | 1.5-2.5 | 2.5-3.0 | 0.3-0.5 |
| Suction Limit | 180 | 108 | 38 |
| Pressure drop (psi) | 4000 | 6670 | 4000 |
| Using time (min) | 10 | 0.6 | 10 |

## DETECTOR

Detector is most important piece of instrumentation required for HPLC for sensitive and continuous monitoring of column effluents.

Detection of solute is difficult when the physical properties of mobile phase and solute are similar. The problem is tackled either by differential measurement of bulk property of solvent and sample or by measurement or sample property not possessed by the mobile phase.

Solvent elimination and pre detector reaction also are employed presently. There are several types of on-stream detectors employed for analysis, important ones are :

1. Refractive Index detector.
2. UV-VIS absorbance detector
3. Fluorescence detector
4. Electrochemical detector
5. Conductivity detector and
6. Flame ionisation detector.

**ANALYSIS**

HPLC is one of the most widely employed analytical technique for the qualitative and quantitative analysis of organic, inorganic and biological species from su-ppp level to percentage level. It is employed in the various fields of science, engineering and technology for research, routine analysis and even quality control.

Its applications in the field of pharmaceutics, forensic sciences, clinical and biomedical studies, food technology, polymer science and environmental pollution analysis are well known.

HPLC is not only employed for laboratory analysis but even for process control, quality control, process stream monitoring in industries and molecular structure analysis, reaction kinetics, size distribution analysis, ionic species analysis, in advanced research laboratories.

The method development of selection of suitable analytical method consist following steps :

1. Selection of mode of chromatographic analysis.
2. Selection of column.
3. Mobil phase selection.
4. Deter selection.
5. Selection of analysis parameters.
6. Selection of other techniques such as gradient elution.

While selecting the mode of analysis, polarity, ionisibility, molecular size distribution, pH and solubility of the sample components should also be considered. Other analysis parameter such as flow rate, sample size, sensitivity etc. depend on the concentration of the sample and time of analysis.

All these parameters are not totally independent and should be optimised after scouting studies. For the analysis of the complex mixture techniques like premixing of solvents, gradient elution, flow programming and multiple column coupling etc. are used.

## APPLICATION IN ENVIRONMENTAL ANALYSIS

HPLC has been effectively employed for analysing polluntants in air, water, waste water and solid wastes. Polyaromatic hydrocarbons (PAH) are carcinogenic compounds formed in low temperature coal burning.

Sensitive and fast analysis of these compounds can be done by employing reverse phase BPC and fluorescence detector.

Various types of pesticides such as organochlorine, organophosphorus, carbamates etc. can be analysed by this instrument.

Other analysis performed by HPLC include pesticides, nitrosoamine in drinking water, carboxylic and sulfonic acids, phenols, aldehydes and ketones in industrial wastewater and various fertilizers and pest killers in agricultural run-off water.

Recent development in HPLC such as high resolution, microbore columns, high speed columns, diaode array detectors, computer aided data handling and capacity to interface with advanced detector like mass-spectrophotometer, atomic absorption spectroscopy and fourier transform infrared spectroscopy have opened new vistas for HPLC. Its universal application would not be surprising.

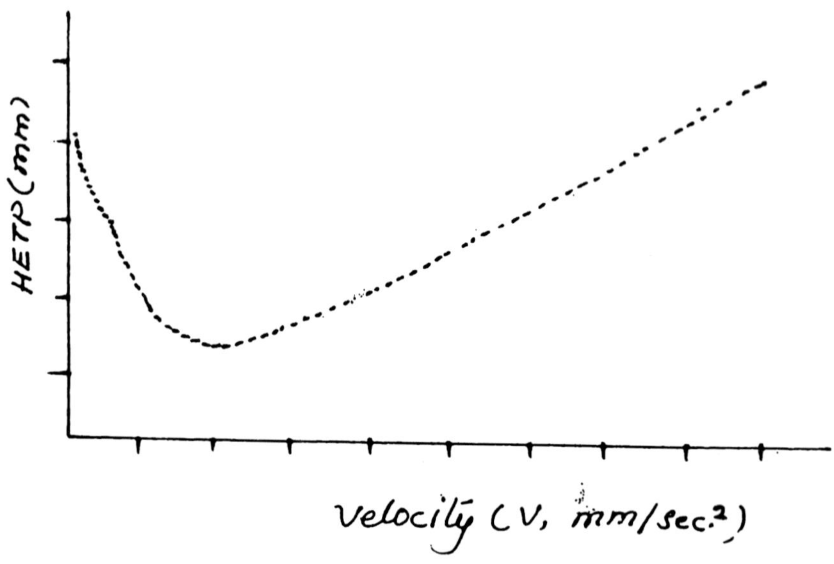


**Fig. 22.1 : A Van Deemeter curve**

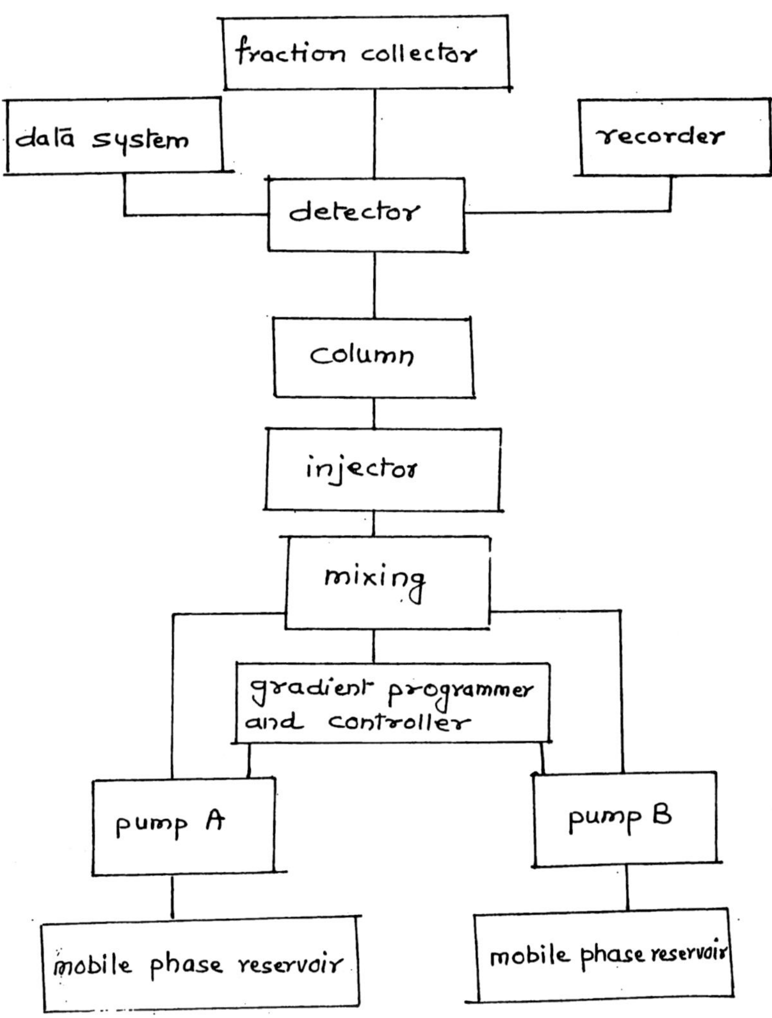


**Fig. 22.2 : Schematic of functional components of HPLC**

## Chapter -23

# GAS CHROMATOGRAPHY MASS SPECTROMETRY (GCMS)

## INTRODUCTION

The environment specialists are concerned with organic pollutants, which exist in the environment some of them are highly toxic and carcinogenic. In fact EPA has designed total priority pollutants of which only 14 are the metals and rest 114 are organic pollutants.

Absolute chemical verification of a particular organic pollutant requires the use of a sophisticated precision analytical instruments called "***gas chromatography and mass spectrometry***".

In mass spectroscopy, organic pollutant is identified by determining its molecular weight and the concentration is known by the abundance of molecular ion.

## TECHNIQUE

In the technique of the organic substance is vaporised and brought under vacuum of $10^{-6}$ to $10^{-7}$ mm to the ion source. In the ion source compartment, the substance molecules are bombarded with electrons of sufficient energy to form positive molecular ions.

These molecular ions are then accelerated and separated by magnetic sector, electric sector or quadrapole mass analyser.

The ion current is then recorded by ultra violet galvanometric recorder or converted to millivolt signal by high independence electrometer amplifier and recorded by the fast response milli volt recorder.

The data storage is done on the magnetic cassettes and recorded by the magnetic tape recorder. The computer campares sample data with the standard data to give line spectrum of molecular species.

The molecular weight is determined by the position of the parent molecular ion and its abundance determines the concentration.

## SAMPLE INJECTION SYSTEM

### Heated Inlet Probe (HIP)

Sample vapour is admitted to ionizaiton chamber from the reservoir and system maintains steady pressure between $10^{-5}$ to $10^{-6}$ mm., that is required to ionise the molecules and to separate the ionised molecules.

When the sample is less volatile, it is heated with the heater to about 200°C and is vaporised before being admitted to the ionisation chamber, as mentioned in Fig. 23.1.

### Direct Insertion Probe (DIP)

This is the most useful probe when samples can be directly injected, vaporised and admitted to the here, reactive gas gets ionised and this is turn ionises the sample by charge transfer.

This needs small transfer energy (20 ev) $ so fragmentation is avoided and parent molecule ion peak is most intense.

The technique is not suitable for structure studies but useful for identification of electron affinity, chlorinated hydrocarbons (pesticides) and polychlorinated biophenyls.

## MASS ANALYSER SYSTEM

a. ***Single focus system***

Magnetic sector single focus mass analyser is the most simple system and achieves the resolution of 300-1000 in smaller systems and 300-7000 in large systems.

One can select lower and intermediate resolution by adjusting source and collector slits, the ions formed in the ion source chamber are accelerated through the slit towards homogenous magnetic field.

By varying the magnetic field, the ions of different m/e can be deflected to the collector and the mass spectrum can be obtained.

The resolution in magnetic sector, MS is determined by the radius of armature and width of source ion chamber.

The standards are thus injected directly and the standard mass spectrum can be obtained and stored on the cassette of the magnetic tape recorder for later comparison with the sample data in a computer.

A small sample with microsyringe is placed in a probe and is inserted via vacuum lock to a position near the ion source.

**Direct sampling from GC**

There are two ways of injecting sample from GC :

1. The different fractions from GC effluent are trapped cooled and brought to liquid state and later injected with DIP.
2. GC can be directly coupled via separators, either jet separator or the membrane separator.

The function of the separator is to

1. remove the carrier gas,
2. concentrate the sample,
3. bring down the pressure of the sample to $10^{-6}$ mm compatible with the mass spectrometer.

The great advantage of MS technique is the requirement of small size. The sample needed in the above three system are given in following Table 23.1.

**Table 23.1 : Sample requirement**

| *Sample system* | *Max. Sample* | *Min. Sample* |
|---|---|---|
| HIP | 100 μg | 1 μg |
| DIP | 10 μg | 0.1 μg |
| GC Sampling | 1 μg | 0.01 μg |

## IONISATION SOURCES

### 1. Electron Impact Source

Electron impact is the most general source used in all the mass spectrometer to produce positive molecular ions.

The electrons of fixed energy (70 ev) are emitted from the surface of the heated rhenium filament and are allowed to colloid with the molecules of the substance to form positive molecular ions.

Although parent molecular ions, fragmentation ions are produced due to breakage of functional bonds.

The fragmentation pattern evolved in E-I spectra is much helpful in determining structure of the molecules.

### 2. Chemical Ionisation Source

Chemical ionisation source gives only added information. In fact, many modern MS have EI/CI combined source so that any source can be switched for the studies.

In CI sample is admitted as a vapour with reactive gas methane and the mixture is bombarded by electrons at lower energies between 14-200V. Slit and that collector slit $S_2$ and is given as :

$$R = \frac{K.r}{S_1+S_2}$$

The radius determines the mass dispersion and the collector slit defines the distance of focused image.

### *b. Double focus system*

The magnetic sector single focus system as a matter of fact does not separate into ion masses but does the momentum separation which broadens the peaks and limits resolution.

To counteract this velocity separation, and additional ions of strong electric field is associated to make it double focusing and the divergent beams are brought to focus.

Thus, the ions with the same mass but different velocity will also be separated and is used in combination with magnetic sector to obtain energy dispension.

Double focusing is achieved in a plane at the border of the magnetic field and the detection is done with photoplate detector.

The purpose of double focussing mass spectrometer is to achieve sufficiently high resolving power from 10,000 to 20,000 and may be useful to separate certain mass doublets or for isotope studies.

### *c. Quadripole system*

This consists of four hyperbolic rods. The opposite rods are electrically connected. The radio frequency component $V_0$ is imposed on the DC voltage component $V_{so}$ that the *f* ratio $V/V_0$ is kept constant and less that one to pass the oscillation ions.

Hence positive molecular ion oscillates between adjacent quadripoles of opposite polarity. When ions of given mass undego stable oscillation they are passed on to the collector.

Ions with unstable oscillations are lost to the quadripoles. Thus the quadripole spectrometers are mass scanned by varying DC voltage V and radio frequency $V_0$ so that $V/V_0$ is kept constant.

The resolution of the quadripole mass spectrometer is between 500-1000. The ability to select wide range of stable oscillation masses by varying law value of $V/V_0$ provides a better method to couple these mass spectrometers with gas chromatographs.

## DETECTORS

### Electron Multiplier Detector

The most electron multiplier detectors have 12-20 dynodes electrically connected through resistive network. The ion beam strikes the dynode (ion current beins $10^{-12}$ to $10^{-18}$ amp) and one shower of electrons are emitted and the process is followed at other dynodes finally the ion current is multiplied to $10^{8}$ times of the initial striking value.

A typical 16 stage copper beryllium detector (2% Be with Cu) has gain of $10^{7}$ at an accelerating voltage of 3-2 Kv between the first and the last dynode at the rate of 200 V/stage.

In most GCMS runs $10^{6}$ gain is used and when higher sensitivity is required the gain 10X and 100X can be obtained to pin point the probable insignificant masses.

## UTTRAVIOLET OSCILLOGRAPHIC RECORDER

Here several galvanometer assemblies each with its own input signal terminal are included into a magnetic block. The current passing through small coil causes deflection in a miniature mirror and reflects the mercury light beam to UV sensitive paper.

The frequency response is fast. The normal galvanometers used are of 1600 cups with a sensitivity of 20 millivolt per mm.

The galvanometer systems with attenuations XI, 1/10, 1/10 X 1/100 are generally used.

Although oscillographic galvanometric recorders are relatively insensitive than electrometer amplifiers, they are useful for lower sensitivity studies.

## MAGNETIC TAPE RECORDERS

For the excess amount of the anticipated data with scan speed exceeding one Sec/decade, faster recording mode is required and here the above recorders are not useful.

For this analog magnetic tape recorder with computer is used. almost all modern of mass spectrometers are completely handled by computers and the storage of sample and standard data on analog magnetic tape recorder is much essential.

For further comparison of standard and sample data is done with the help of the computer. The principal application of tape recording for low resolution GCMS is with spectra that are to be processed by a computer.

A small analog tape unit can acquire the data at any reasonable scan rate and resolution for further processing with an in-house computer.

## VACUUM SYSTEM

The fore pump used for high vacuum can be rated 1t 150-200 liters per minute and can produce the vacuum upto $10^{-12}$ mm.

The fore pump is used with oil diffusion pump is 0.2 to 0.5 mm and beyond this value the diffusion pump is automatically switched off in the oil diffusion pump, the fluid is boiled up in the chimney and directed down the jet, in high velocity stream of vapour. The momentum is transferred to gas molecules which are brought at the bottom of the forepump.

The pump vapours condense on the water cooled walls and return to the boiler. The successive stages can attain vacuum of $10^{-6}$ mm with cooling waters.

The successive stages can attain vacuum of $10^{-6}$ mm with cooling waters. The vacuum of $10^{-8}$ to $10^{-9}$ mm can be attained with liquid nitrogen cold provided with most of the mass spectrometers.

The diffusion pumps are rated at 150-500 litres per second. High vacuum measurement is done with hot filament ionisation gauges. This produces positive current of ionised gas molecules. The current is proportion to vacuum created and the scale of the gauge is calibrated in mm.

## OPERATION

1. Molecular ion : The molecular ion is formed by the loss of one electron from the molecule and is known as ***molecular ion***. The peak corresponding to it, is ***molecular peak***. The examination of peak position gives the molecular weight of the molecule under investigation.

2. The peak corresponding to molecular ion is even mass except when nitrogen is associated, the odd mass is obtained.

3. The mass marking is being done by injecting standard per fluoro compounds such as per fluoro-tributylamine (PTBA) having mass peaks at 69, 2990 and 503.

4. The ions formed in intermolecular process such as in $Cl_2$ give (M +1) + peak for the molecular ions and is useful in the study of the chlorinated hydrocarbons.

5. The data presentation relative abundance of each peak restricting to five peaks for each mass can be done as shown in the Table 23.2.

**Table 23.2 : Relative abundance of various peaks**

| M/l | abundance (%) |
|---|---|
| 26 | 7.2 |
| 27 | 25.1 |
| 51 | 10.1 |
| 52 | 7.2 |
| 78 | 100 M+ |

6. **Framgentation pattern**

Though intense molecular ion peak from $M^+$ is obtained several bonds of substance are broken by high energy electron bombardment in electron impact giving rise to M/e seaks for less that $M^+$.

Thus, for naphthalene ($C_{10}H_8$=128) though we get M+ intense peak at 128, the peak at M-$CH_3$ when $CH_3$ bond is broken can be obtained at a distance ($CH_3$=15) from molecular ion. In saturated hydrocarbons with even electrons $CH_3$ bond breakage can be easily observed.

7. Normally there are no ions between 18($H_2O$)+ and 24($C_2$)+, nor between 2($H_2^+$) and 12 (C)+. The pair of peaks may be obtained due to presence of air at M/c=28 ($N_{2+}$) and M/e=32 ($O_2^+$) in a ration of 5:1 intensity separation.

## APPLICATION IN ENVIRONMENTAL ANALYSIS

This instrument provides a sensitive technique for identification and quantitation of environmental pollutants to a nanogram level. Some of the PAH with M/e is given in the Table 23.3

**Table 23.3: Polycvclic aromatic hydrocarbon (PAH)**

| *Name of PAH* | *mol. formula* | *m/e* |
|---|---|---|
| Napthalene | $C_{10}H_8$ | 128 |
| Acenapthalene | $C_{12}H_{10}$ | 154 |
| Phananthrene | $C_{14}H_{10}$ | 178 |
| Pyrene $C_{16}H_{10}$ | 202 | |
| Fluoranthene | $C_{16}H_{10}$ | 202 |
| Chrysene | $C_{18}H_{12}$ | 228 |
| Benzo(a) pyrene | $C_{20}H_{12}$ | 252 |
| Anthrathrene | $C_{22}H_{14}$ | 276 |

**2. Pesticides :**

The molecular ion peaks for come of the pesticides are given in Table 23.4.

**Table 23.4 : Molecular ion peaks of pesticides**

| *Pesticides* | *m/e for molecular ion* |
|---|---|
| BHC | 181 |
| DDT | 235 |
| DDE | 246 |
| Aldrin | 261 |
| Dieldrin | 378 |
| Endrin | 378 |
| Heptachlor | 370 |
| Heptachlor epoxide | 386 |
| I - Chloradane | 406 |

### 3. Polychlorinated Biphenyls (PCB)

For molecular ion peaks for some of the PCB compounds are given in Table 23.5.

**Table 23.5 : Molecular peaks of some PCB**

| *Compounds* | *M/e for M+* |
|---|---|
| Dichlorobiphenyl | 222 |
| Trichlorobiphenyl | 256 |
| Tetrachlorobiphenyl | 290 |
| Pentachlorobiphenyl | 324 |
| Hexachlorobiphenyl | 353 |
| Octachlorophiphenyl | 426 |

With higher order biphenyl. The presence of chlorine atmosphere molecule increases and this increases the complexity of mass spectra.

### 4. Chlorophenols :

For chlorophenols m/e are given in Table 23.6.

**Table 23.6 : Molecular peaks of some chlorophenols**

| *Compounds* | *M/e* |
|---|---|
| Monochlorophenol | 128 |
| Dichlorophenol | 162 |
| Trichlorophenol | 196 |
| Tetrachlorophenol | 230 |
| Pentachlorophenol | 264 |

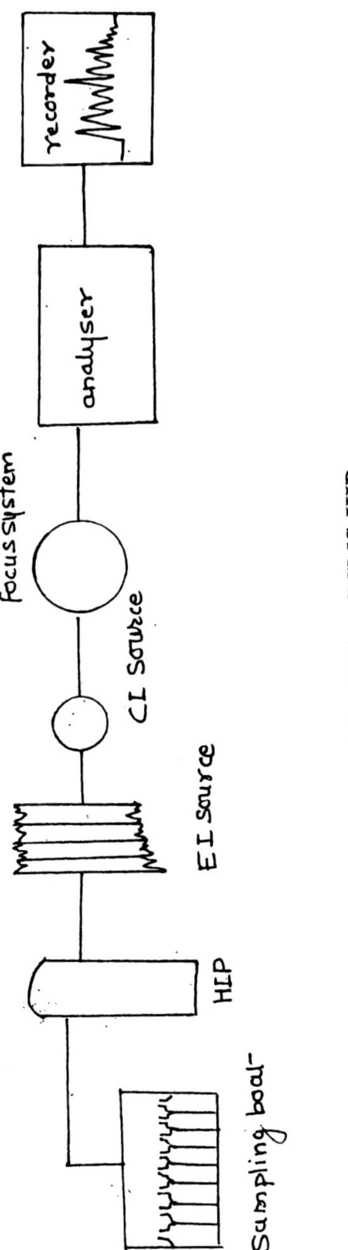


**Fig. 23.1 : GCMS-HIP**

## Chapter -24

# ION-SELECTIVE ELECTRODES

## INTRODUCTION

Ion-selective electrodes being to a class of electrochemical sensors. Their use involves a special care of direct potentimetry.

When these generate a millivolt potential related to the concentration or more correctly the activity of a given ion and over a wide range the potential is linearly related to log (activity).

These are the devices which give directly an electrical output related to composition of the sample solution.

Historically, the oldest best known & most widely used member of ion selective electrodes family is the glass electrode, which responds to hydrogen ion activity & it remained the only sensing electrode until mid 1960's.

But the renewed interest in potentiometry has led to the commercial availability of a range of ion selective electrodes which open the door to direct & indirect measurement of dozens of chemicals species that are of vital importance in the water quality control.

## BASIC PRINCIPLE

### a) Electrodes : General Description

The classical ion-selective electrode consists of a tube made of a good electrical insulator which is closed at its lower end by a sensing membrane.

Within the tube there is a filling solution known as internal filling solution containing a fixed concentration of the ion to which, the membrane is sensitive.

A stable electrical contact is made with the internal filling solution by dipping a silver wire coated with silver chloride (known as internal surface electrode).

**b) Working Mechanism**

When the ion selective electrode is placed in the sample solution, the inner filling solution is in contact with the inner face of the membrane while the sample solution contacts the outerface.

If the concentration of the sample solution is greater than that of the filling solution, ions migrate inward with their associated charges, thus, changing the potential across the membrane.

An equilibrium potential soon develops which just prevents further ingress of ions. Migration takes place outward if the sample is less concentrated than the inner solution.

The number of ions transported is so small that the change in sample concentration is infinite-simal. The magnitude of potential developed across the membrane depends upon the activity of the species being determined in the sample solution. However, the potential of an ion-selective electrode cannot be measured by itself.

It can only be measured in conjunction with an external reference electrode whose potential relative to the solution is intended to be constant regardless of composition.

The relation between electrode potential and activity is given by ***Nernst equation***:-

$$E = E^{o} + 2.303\frac{RT}{nF}\log a$$

Where

$E$ = the electrode potential

$E_0$ = a constant for a given combination of ion selective & reference electrode

$R$ = the gas constant

T = absolute temperature

n = ionic change of the analyte ion

F = the Faraday constant

a = activity of the analyte ion

## OPERATION

The fundamental Nernst's equation gives information about the activity of the ion being measured. The data in terms of activity is acceptable where one is interested in the determination of rates of reaction.

For example, in case of water hardness, it is the activity of divalent cation that determine the extent of scale formation & not the total concentration of calcium and magnesium. However, in most of the pollution studies, data in terms of concentration is requested. Thus, in order to trick the electrodes is giving information in terms of concentration the relationship-

$$a = f.c.$$

Where,

f = activity coefficient of the analyte ion

c = concentration of the analyte ion is moles per litre.

But the absolute determination of activity coefficient which depends upon the total ionic composition of solution, is a tedious & time consuming task.

Therefore, the other alternative is to maintain it constant. Hence following equation is obtained which indicates that potential measurement are linear with respect to concentration-

$$E = k + \log.C$$

Where,

E = a constant (including all constants) for the given experimental conditions.

S = Slope factor, positive for cation and negative for anions and proportional to absolute temperature

(59.16 mV for monovalent ions & 29.58 mV for divalent ions at 25°C)

In order to minimise uncertainties due to variation in the total ionic composition, samples and standards should be of similar ionic strength. This requirement is usually met by either-

1. Preparing standards for solution of similar composition to those which are to be analysed, but containing essentially negligible amount of ion of interest or,

2. Adding a non interfering electrolyte to both sample as well as standard solutions. In this case the total composition is effectively governed by the added iner telectrolyte to which the ion-selective electrode does not respond.

Both methods are in common analytical use and the choice between them is best made by the analyst after considering the particular requirements of the analysis. On balance method is preferable because it provides a more positive compensation for the possible effects of ionic strength variations.

## INSTRUMENTATION

Excluding the requirements for sample pre-treatment (when necessary), constant temperature bath & magnetic stirrer the following equipments are needed in order to obtain quantitative results with this analytical techniques.

### 1. Potential measuring instrument

The nature of response of an ion selective electrode to changing concentration consists of an output potential whose magnitude changes by about 60 mV for a univalent and 30 mV for a divalent ion if the concentration changes by a factor of 10.

The potential measuring instrument used with these elecrodes must therefore be capable of measuring such potentials.

Secondly, because the Nernst's equation strictly applies only to equilibrium conditions at the membrane, i.e. when no net chemical reactions occur it is necessary to use instruments with high input-impendence to measure the electrode potentials.

If this condition is not met, large currents, might be drawn for ion-selective electrode and reference electrode & these currents would tend to cause departures from equilibrium membrane conditions.

Another reason for using high impendence meters is the extremely high resistance of some types of ion-selective electrodes.

For typical range of interest of ionic activity or concentration (e.g. $10^{-1}$ to $10^{-6}$ M) approximately 300 mV for univalent and 150 mV for divalent ions are suitable maximum values of full scale deflection of the meter scale.

To permit both union and cation sensitive electrodes to be used, the meter scale should be reversible, a polarity-reversing switch is therefore necessary on commercially available meters for the use with ion-selective electrodes.

**2. Reference electrode**

The potential of an ion-selective electrode is measured against that of a reference electrode. The principal types that can be used for this purpose are-

1. the calomel electrode
2. the mercurous sulphate electrode and,
3. the silver/ silver chloride electrode

However the choice of a suitable reference electrode for use is analysis is important if satisfactory results are to be obtained. The selection of a suitable reference electrode for use in water analysis will largely depend on the particular analytical situation. For instance in the determination of chloride using chloride selective electrode whose filling solution contains chloride ions will not be satisfactory, as chloride ions will diffuse through the liquid function of the reference electrode into the sample whose chloride content is being determined. In that case a double junction reference electrode should be used. In on-line analysis, however, the use of a single junction reference electrode for the same determination may well be satisfactory because it is often possible to mount the reference electrode in a specially constructed cell through which the sample flows.

## MEASUREMENT TECHNIQUES

Various measurement technique can be used during the use of ion selective electrode analysis.

### a) Direct method

This method is useful for analysing a large number of sample as rapid measurements are possible over any to decade of concentration. The concentration of free species in sample solution can be read directly from logarithmic scale of ion-meter after calibrating the scale with known standards which differ in concentration by a decade.

### b) Calibration curve method

This method is most commonly used. Standards are prepared by serial dilution. The electrodes are placed in the standardising solutions & potentials are recorded. Then a calibration curve is prepared by plotting the electrode potentials are recorded and the corresponding concentrations are read from the calibration curve.

### c) Standard addition method

This method is convenient for measuring occasional sample because it eliminates the time required to prepare a calibration curve.

The method consists of taking an initial potential reading by dipping electrodes in the sample solution & then taking a second reading after addition a known volume of standard solution.

From the resulting change in potential the original total concentration of analyte ion can be calculated. Direct reading standard addition scales are also now available on ion meters.

### d) Sample addition method

Sample addition method is adopted when standards for species to be measured are difficult to prepare.

In this method, the sample is added to a reagent of known concentration that reacts with the sample. The change in the potential reading before and after sample addition is used to measure to concentration of the analyte ions.

Since an electrode sensing the reagent species is used, this technique is also useful in situations where no electrode exists for the species to be measured.

## FACTORS AFFECTING ELECTRODE MEASUREMENTS

### 1. Effect of pH

The pH of the sample solution changes the chemical environment of the analyte ions e.g. at low pH values, $F^-$ ions are present as HF which are not sensed by the fluoride electrode.

Similarly in case of ammonium ion measurement, it necessary to convert $NH_4^+$ in $NH_3$, which is only sensed by the ammonia electrode.

Thus in such cases an appropriate pH adjuster should be added to sample to set up a reliable ion-selective electrode analysis.

### 2. Presence of complexing agent

A number of ions in water analysis can be present in more than one form such as complexes with other ions or molecules.

Electrodes can not sense bound or complexed ions. This property of electrodes can be useful when it is required to distinguish between the different forms in which a given determined may be present in sample. e.g. a copper ion electrode will distinguish between free and bound $Cu^{2+}$ ions.

However, in other situations, when it is the total concentration of a determined that is of interest, the property maybe regarded a disadvantageous.

When complexing agent is present in small quantity, a reagent is added that frees the ion of interest. However, in presence of large excess of complexing agent the standard addition method of applied.

### 3. Temperature Effects

It is necessary to control tempeature of the solution, because the slope factor in the Nernst's equation is tempereature dependent.

Some types of magnetic stirrer generate considerable heat when in operation, and this may cause the temperature of sample to change sufficiently. This prevents the electrode from rapidly attaining a steady potential, but a piece of thermally insulting material placed between the beaker & magnetic stirrer surface assists in reaching a steady potential.

For most routine application control to within ± 2°C of room temperature is adequate. For most precise measurements control to within ± 0.5°C is desirable and for other thermostatic devices are necessary.

The other alternative is to measure sample temperature independently an appropriate correction during commutation of results. But this method has the disadvantage that delay is inherent before the result of determination can be obtained.

## CLASSIFICATION OF ELECTRODES

In view of the large number of commercially available electrodes, it is necessary to group them according to the physical state of sensitive membrane.

### a) Glass electrodes

In these electrodes, the membrane is a thin layer of glass, a familiar example being the pH electrode which has a high sensitivity for hydrogen ions.

### b) Solid-state electrodes

In these electeodes, the solid state membranes are made of either a single crystal for example the fluoride electrode which uses a lanthanum fluoride crystal or the membrane consists of a pallet of insoluble salt.

The commercially available halide electrodes contain membrane which is composed of a mixed $AgX/Ag_2S$ (where x = chloride, bromide, iodide) pallet & the metal ion electrodes contain membrane composed of a mixed $Ag_2S/MS$ (where M = copper, cadmium, lead etc.) pellet.

### c) Liquid ion exchange electrodes

In there, a liquid ion exchanger (usually an organic compound dissolved in a non polar solvent immiscible with water) is used.

The analyte ionic species can combine with ion exchanger & electrolytic contact combine with ion exchanger and electrolytic contact between this and test solution is maintained through an inert porous membrane which allow the ion exchanger to pass through but does not allow water to enter the electrode. Examples of this type are the electrodes available for nitrate and divalent cations.

**d) Gas-transfer membrane electrodes**

The electrodes belonging to this class are not strictly ion electrodes because they respond to dissolve molecular species.

It consist of a hydrophobic membrane which separates a reference solution of fixed concentration from a test solution in which it may vary.

Diffusion of the species through the membrane occurs in either direction until the concentration of the species on both sides of the membrane are equal, this process causes a pH range in the reference solution, which is detected by a suitable combination of glass and references electrode.

The ammonia electrode is the best known example of this type is environmental studies.

## PERFORMANCE CHARACTERISTICS

**1. Response time**

The response time of ion selective electrode is the speed at which it responds (electrode potential reaches a steady state value).

The response time depends upon the type of electrode and concentration of ions under measurement. The response time of solid state electrode is generally of the order of 5-10 seconds.

On the other hand, the response time of liquid ion exchanger electrode is longer, usually several seconds or even few minutes at low concentration.

In the presence of some interfering ions, the lengthening of response time has been observed. The response time is much shorter with defined stirring of solution than in unstirred solution because the establishment of steady membrane potential also depends upon the transport of ions across the interface.

**2. Potential drift**

The potential drift of ion selective electrode is connected with changes in the surface structure of solid membrane electrodes due to contact with the electrolyte and due to dissolution of the ion exchanger of liquid membrane electrodes.

All electrodes show drift with time and therefore need recalibration periodically with standardising solutions.

If the reproducibility of the order of 5% is acceptable, solid state electrodes can be restandardised as frequently as once a day, but liquid ion exchange electrodes need recalibration every few hours.

Additionally, some electrodes exhibit a slight concentration hysteresis, therefore, in a series of measurements, it is preferable house electrodes in the direction of rising concentration.

**3. Life time**

The life time of electrode is the period, determined by the rate at which the liquid ion exchanger leaks through the membrane in liquid membrane electrodes and in solid state electrodes, the time required for crystal surface or the pressed pellet to become chemically fouled.

Information on life time is very scarce since this markedly depends upon conditions of use. Good electrodes should provide many months or years of service. However, if the electrode respond drops off rejuvenation of electrodes is possible by refilling the liquid internals.

**4. Sensitivity**

The useful working range of the ion-selective electrodes is known as ***sensitivity*** which is effectively governed by the properties of the membrane.

The working range of electrode is determined from the concentration (mg/liter) to which an electode responds in a theoretical fashion.

In solid state electrodes, the lower practical concentration depends upon the solubility of the membrane material. For liquid ion exchange electrodes, variation in the solubility of the organic liquid in aqueous phase affect both the upper & lower working limits of the electrodes.

The only exceptions to these type of limitations are the glass pH and sodium electrodes, where the ultimate limit is set either by the presence of interferences or the absence of a sufficient number of ions in the sample fails to provide for a thermodynamic equilibrium within a reasonable time.

## 5. Selectivity

Although intended to be specific to a single ionic species, most electrodes respond in varying degree to presence of other ions.

The effect of one ionic species on the potential developed by an electrode selective to a different species in given in following equation:-

$$E = K + S \log (C_i + K_{ij} C_j^{\,ni/nj})$$

Where

$n_i$, $n_j$ = Charges of the ionic species

$c_i$, $c_j$ = concentration of ionic species

$K_{ij}$ = Selectivity constant of the electrode for the species i in the presence of species j.

The total effect of all other ions can be obtained by summing the values of $K_{ij}$ terms for the ionic species of interest.

Value of selectivity constants are after quoted by the manufacturer's as an indications of the electrode's selectivity. The knowledge of the values of selectivity constant permit the assessment of importance of potential interferences in a given analytical system.

When the value of $K_{ij}$ are less than $10^{-3}$, the electrode response for species i is, for practical purposes, independent of the concentration of j. However, where serve interferences are likely to countered, there are two practical courses of action.

Either the electrode must be calibrated in the presence of interfering ion or the interference must be eliminated by some other means such as buffering, precipitation or by addition of complexing agent.

### 6. Precision & Accuracy

As the ion selective electrode analysis is a logarithmic device the electrode gives a constant precision throughout its dynamic range.

A measurement of 0.01 ppm will have the same precision of 1000 ppm. Regarding the accuracy of measurement of 0.01 ppm will have the same precision of 1000 ppm. Regarding the accuracy of measurement by ion selective electrodes, it depends on the choice of reference electrodes, temperature control and presence of interferences. But the relative error in the given measuring system is dependent only on the absolute error in the potential measurement.

An error of 0.1 mV leads to 0.4% error in the concentration measurement of monovalent ions and 0.8% in case of divalent ions.

## APPLICATION IN ENVIRONMENTAL ANALYSIS

A list of the commercially available electrodes most likely to be interest in environmental analysis together with their characteristics are given in Table 24.1. The initial applications of these ion selective electrodes have been made in the laboratory.

Table 24.2 gives a survey of the application of some ion selective electrodes in water analysis. About the other ion selective electrodes, although papers dealing specifically with the use for water analysis are very scarce.

## SIGNIFICANCE

1. Do not bother for turbidity or colour of the sample.
2. Small volumes of samples can be analysed without loss.
3. Measure species without upsetting chemical equilibrium.
4. Measurements are independent of volume sample.
5. Concentration can be read directly on specific ion meter.
6. Calibration curves can cover even six decades of concentration.
7. Electrodes are sensitive down to parts per billion (ppb).

8. Do not require extensive pre-treatment of the sample.
9. Relatively rapid response time.
10. No need to have special and extensive training for operation.

**Table 24.1 : Ion selective electrodes in Environmental analysis**

| *Ion/molecule* | *Types of membrane** | *pH range* | *Lower limit of detection (moles/lt.)* | *Principal Interferences (ions)* |
|---|---|---|---|---|
| *1* | *2* | *3* | *4* | *5* |
| Ammonia | GT | 7-14 | $10^{-6}$ | Amines |
| Bromide | SS | 2-12 | $5\times10^{-6}$ | Iodide, cyanide, sulphide. |
| Cadmium | SS | 2-12 | $10^{-7}$ | Silver, mercury Ferrous |
| Calcium | LM | 5-10 | $10^{-5}$ | Zinc, ferrous, lead, copper, nickle |
| Copper | SS | 2-10 | $10^{-8}$ | Silver, mercury Ferrous |
| Chlorine | SS | 4-5 | $10^{-7}$ | Strong oxidants |
| Cyanide | SS | 12-214 | $10^{-6}$ | Sulphide, iodide |
| Fluoride | SS | 5-8 | $10^{-6}$ | OH |
| Hydrogen | GM | 0-14 | $10^{-14}$ | None below pH 13.0 |

| *1* | *2* | *3* | *4* | *5* |
|---|---|---|---|---|
| Iodide | SS | 2-12 | $5\infty10^{-8}$ | Cyanide, sulphide |
| Lead | SS | 2-10 | $10^{-7}$ | Silver, mercury Ferrous |
| Nitrate | LM | 3-10 | $5\infty10^{-5}$ | Iodide, bromide, Nitrite, bicarbonate |
| Potassium | GM | 3-12 | $10^{-6}$ | Hydrogen, sodium, silver |
| Silver | SS | 2-10 | $10^{-17}$ | Mercury |
| Sulphide | SS | 12-14 | $10^{-7}$ | Mercury |
| Sodium | GM | 3-12 | $10^{-6}$ | $A^{-6}$ Hydrogen, potassium, silver |
| Water | LM | 5-8 | $10^{-8}$ | Hydrogen, ferrous, copper, lead |

* Electrodes classification

SS - Solid state

GM - Glass membrane

GT - Gas Transfer

LM - Liquid membrane

**Table 24.2 : Application of ion selective electrodes in Environmental studies**

| *Constituent determined (mg/lt.)* | *Concentration range* | *Matrix* | *Remarks* |
|---|---|---|---|
| *1* | *2* | *3* | *4* |
| Ammonia | 0.02-2.0 ammonia nitrogen | Surface water sewage saline water | Filtered & added NaOH to adjust pH for $NH_3$ estimation |
| Chloride | 0.1-100 | Boiler feed water | 5M Nitriacid added to sample & applied differential potentiometry |
| Cyanide | down to 2.0 | Industrial Wastewater | Free & complex Cyanide determined |
| Calcium | 20-800 | fresh water | Buffer added to provide constant pH & ionic strength |
| Copper | down to 1 μg/lt. | Sea water | Calibration with sea water in the absence of complexing agent |

| *1* | *2* | *3* | *4* |
|---|---|---|---|
| Fluoride | 20 μg/lt. | fresh & sea water | Direct or potentio-metric & known addition technique |
| Nitrate | 5-20 | Drinking water | buffer solution used to prevent Cl & $HCO_3^-$ interfer-ences |
| Sodium | 1-50 μg/lt | Boiler feed water | for power station |
| Sulphite | 0.003-3.0 | fresh water | Direct & titrametric |
| Sulphate | 1-10 | Natural water | Dilute 1+1 with methanol |

## Chapter -25

# ELECTROPHORESIS

## INTRODUCTION AND BASIC PRINCIPLE

The process electro-phoresis can be defined as-

***The migration of charged molecule or ions under the influence of electric field is called electrophoresis.***

The charged molecules or ions when placed in the electric field migrate towards one or the other electrode depending on the nature of the charge. If a direct current is passed through a solution having positive ionic species and negative ionic species, the cation species (+) will move towards the cathode and the anionic species will towards anode.

This property of electrically charged substance thus can be used to bring about the separation of closely related compounds.

## TYPES OF ELECTROFORESIS

There are several types of electrophoresis but commonly used methods are as follows:-

a) ***Boundary electrophoresis***

This is moving type, in which ions move as a boundary in solution.

b) ***Zonary electrophoresis***

This is a diffusion type electrophoresis, where the ions diffuse through a supporting medium like filter paper, starch, agar or acrylamide gels. This method is commonly used in the laboratories. The compounds separate distinctly into zones or bands, from which, they can be isolated.

## FACTORS INFLUENCING ELECTROPHORESIS

There are several factors which influence the separation of compounds during electrophoresis-

a) The shape and size of molecules.

b) The pH of the solution.

c) The amount of current applied.

d) The time for which, the current is run.

e) The characteristics of separating medium.

f) The electro-endomosis phenomenon.

## ELECTROPHORETIC INSTRUMENTS

A number of designs are manufactured by different manufacturers, but the basic principle remains the same as given below:

There are two tanks, each provided with a platinum electrodes. The tanks are filled with the buffer solution and the two tanks are made in connection with earth other through a separating medium, paper or agar gel/starch gel.

The contact between separating medium and thus buffer is usually made by soaked filter paper wicks instead of dipping the two ends into the buffer.

The electrodes are connected through a power back. If the experiments is to run for a longer period, arrangement is provided for its cooling by circulating water or any other constant around the unit.

## EXPERIMENTAL PROCEDURE

1. Separation of serum protein by paper electrophoresis.
2. Separation of serum protein by agar gelatin electrophoresis.
3. Prepare a buffer solution of 0.075 m barbitone and HCl, having pH 8.6 and fill the tanks. Prepare 1% solution.

# ELECTROPHORESIS APPARATUS

## AREAS OF APPLICATION

The apparatus is mainly used in the study of carbohydrates, protein, amino acids separation of serum protein, which is effectively done by it.

### Principle

Charged ions or charged molecules placed in a electric field migrates towards one of the electrodes. This property can be used to bring about the separation of closely related compounds.

A strip of filters paper (Whatman's No. 1 or 3) moistened with a suitable buffer solution is supported on a bridge and its ends are kept dipped into reservoirs by buffer solution.

Electrodes are kept immersed in these solution, which are connected to a D.C. voltage source, thereby setting up potential gradient along the length of the paper. A drop of the solution under analysis is placed on the paper and a current is passed through it for few hours.

The substances constituting the sample under analysis tend to migrate at different speeds and at the end of the electrophoretic run they get separated from one another (Fig. 25.1).

### Operating instructions

Connect the pair of leads to the power supply on the one hand and the migration chamber on the other by inserting the plug of each lead in the corresponding sockets.

Now, the instrument can be operated under any one of the following mode of operation:-

a) Through voltage mode and/or

b) through current mode of operation.

### Voltage mode of operation

1. At the switch marked 'Mode' at the position voltage.
2. The potentiometer marked 'Volts' is turned to the extreme anti-clock wise position.

3. The HT toggle switch is kept in 'OFF' position.

4. Now, the main switch is turn to 'ON' position, when the indicator lamp on the panel glows.

5. Three positioned VOLTAGE RANGE switch is kept in position corresponding to the range in which the selected voltage lines.

   The required voltage is adjusted by turning the volts knob clockwise while setting the voltage the '**PRESS FOR VOLTS**' switch is kept in pressed position.

   The magnitude of the voltage depends on the duration in which one wants to complete the operation. Generally about 4-6 volts per centimetres of paper length giving a current of 1.0 to 1.5 M.A. per 4 c.m. width strip for a 16 hrs. run.

   The meter is marked in **'VOLTS'** as well as **'M.A.'**

   Note the switch 'Press for Volts' in the normal position (without pressing) indicates the current passing through the paper strips.

6. The HT toggle switch is thrown to 'ON' position.

## CURRENT MODE OF OPERATION

1. Keep the switch marked MODE at the position marked current

2. Turn the potentionmeter knob marked 'CURRENT' to the extreme left (anti-clock wise position).

3. Put the HT toggle switch of **OFF** position.

4. Now, switch on the mains switch to ON position, when the bulb glows.

5. Keep the voltage range switch in 0-100 V range.

6. Switch on HT toggle switch to 'ON' position.

7. Adjust the desired current is not achieved by turning the knob clockwise or anti-clock wise completely and read the scale on the panel meter on the lower side.

8. If the desired current is not achieved by turning the knob clockwise completely, then change the voltage range switch to higher range.

9. For the first few runs, keep the **'POLARITY'** switch to 'NORMAL' position but for later runs, keep it in the 'REVERSE' position.

10. When the run is over, the HT switch put off.

11. Remove the lid and take out the paper strip.

12. Dry the strip for 30 min. at 100°C in an oven.

13. Stain the strip for 10 min. and dry the strip in air.

## SERUM PROTEIN SEPARATION

### a) Buffers

I ***Barbitone buffer (pH 8.6)***

Prepare a buffer of barbitone pH 8.6 by dissolving 1.84 gms of barbitone and 10.30 gms of sodium barbitone in one lt. dist. water. This will give ionic strength = 0.05 This buffer is used for 7 hrs. run. For overnight run (16-18 hrs), barbitone buffer of ionic strength 0.075 can be prepared by dissolving barbitone (2.76 gms) and basbitone (sodium salt) (15.4 gms) in one lt. distilled water.

II a) ***Sodium tetraborate (pH 8.6)***

It can be prepared by dissolving 8.80 gms of sodium tetraborate and 4.65 gms of Boric acid in one litre of distilled water.

b) ***Sodium tetraborate (pH 9.0)***

It can be prepared by dissolving 7.63 gms of sodium tetraborate and 0.62 gms. of boric acid in one litre of distilled water.

III. ***Sodium phosphate (pH 7.4)***

It can be prepared by dissolving 0.60 gms. of sodium di-hydrogen phosphate (hydrated) and 2.20 gms of di-sodium hydrogen phosphate in one litre of distilled water.

Fill any of the above buffer in the side tanks of the electrophoresis chamber.

**b. Whatman filter paper**

Generally, Whatman filter paper No. 1 is used for analysis purpose, whereas Whatman filter paper No. 3 is used for preparative purposes.

i) Cut 6 c.m. wide strip from the filter paper of the required length (according to the size of chamber).

ii) Draw a pencil line across the paper about one third of the distance from one end of the paper.

iii) Dip the strip in the buffer of choice and the blot it lightly between two pieces of filter paper.

iv) Stretch the strip on the electrophoresis tank in such a way that the pencil line faces the cathode end.

v) See that two ends of the strips are dipped into buffer solution.

**c. Spotting the sample**

Now, apply the sample along the pencil line with the help of a micropipettes at a distance of 1c.m. each.

Take the help of the Ruler guide ( provide along with the apparatus). Care should be taken in spotting the sample. The volume of the spot can be double, it the minor component of the sample are also to be detected.

**d. Running of electric current**

i) ***For 7 hours run***

In barbitone buffer (pH 8.6) 12 MA current is passed for 6 c.m. wide paper strip.

ii) ***For overnight run***

In barbitone buffer 6 MA current is passed for 6 cm wide paper strip.

The electrolysis is carried out at the room temperature. After the run is over (7 hours or overnight as the case may be) switch off the current.

**e. Drying the strip**

i) Remove the strip from the electrophoresis tank with the help of blunt forces.

ii) Dry the step for 10 min. in an over at 100°C in order to prepare it for staining. Proteins gets denatured at the temperature.

**f. Staining the strip for protein components**

i) Generally, Bromophenol blue is used for staining Emerge the dried Strip in the bromophenol blue solution (1% W/V) in 95% (V/V) ethyl alcohol saturated with mercury chloride for 10 minutes.

ii) Wash the strip in several rinses of 1% v/v acetic acid so that the paper background becomes relatively clear.

iii) Wash the strip finally twice in methyl alcohol.

iv) Dry the strip in open air (NOT in OVEN).

**OTHER STAINS**

a) ***Azo-carmine -B***

This is prepared by dissolving 0.1% W/V in 50% V/V methyl alcohol containing 10% v/v acetic acid.

The dried strip is dipped in this solution for 10 minutes and then rinsed thoroughly in 10% v/v acetic acid and finally in methyl alcohol.

b) ***Light green***

The solution is prepared by dissolving light green 0.5 w/v in 25 percent v/v ethyl alcohol containing 5% v/v acetic acid.

The strip is immersed in this solution for 5-10 minutes and the excess of dye is removed by washing the strip in 2% acetic acid.

c) ***Nepthalene black 12-B***

This stain is prepared by dissolving the dye 1% w/v in 10% v/v acetic acid.

The strip is immersed in this solution for 10 minutes and excess strain is removed by washing the strip in 10% v/v acetic acid and methyl alcohol with final washing in methyl alcohol.

### g. Staining the strip for lipoproteins

1. ***Sudan-black-B***

An over saturated solution of this dye is prepared in warm 50% ethyl alcohol. It is, then, cool down to room temperature and dried.

The dried strip is immersed in this solution for few hours and then washed in 40% v/v ethyl alcohol several time to clear background of the paper.

2. ***Acetylated Sudan black-B***

The stain is prepared by dissolving 500 mg. of dye in 50 ml. of warm alcohol. The solution is cooled down and dried.

This dye is used prior to electrolysis i.e. the serum proteins are stained with this dye before spotting the serum on the strip. The dye is mixed with serum in the ratio of 1 dye : 10 serum. The mixture is then, allowed to stand an hour or so then filtered in order to remove the excess particles of the dye and any precipitated protein. The mixture is spotted on the strip using 10 μl per centimetre width of paper and the electrophoresis is carried out for a short period (4-6 hours).

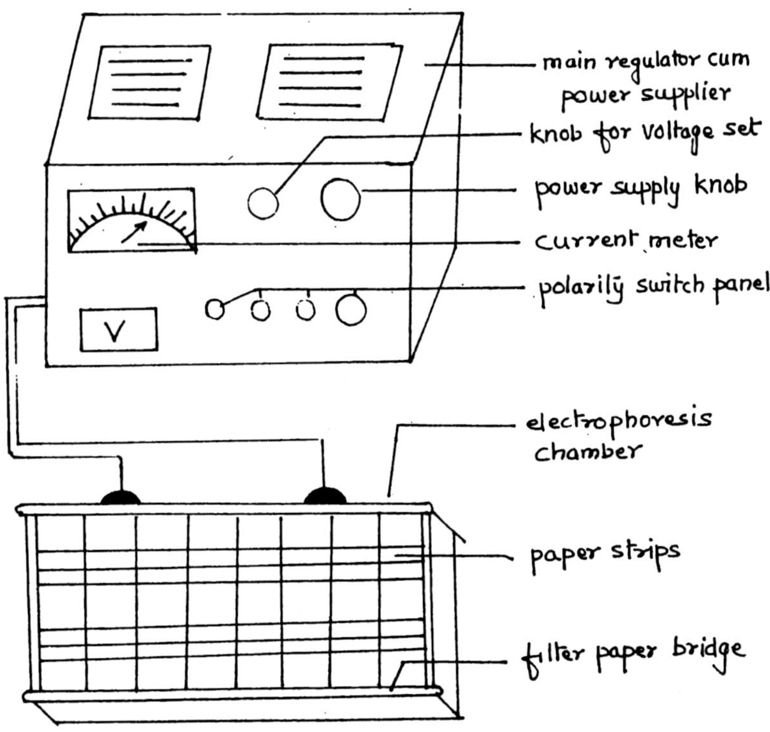


**Fig. 25.1 : Electrophoresis Set-up**

## Chapter -26

# POLAROGRAPHIC ANALYSIS

## INTRODUCTION

The polarographic method was first described by Heyrosky of Charles University in Prague in 1922, using the electrochemical laws postulated by Nernst. He developed principles and instrumentation for the current voltage relationships and called the method Polarography.

Most of the concepts & design factors have remained unchanged to the present time. The earliest polarographic instrument were a recording type, using polarographic paper to trace the polarograms. The electrode systems were nearly identical to those in use today.

## APPLICATION

The polarographic method with its associated procedures is one of the most sensitive analytical techniques available. It is widely used in research and in routine analysis.

## PRINCIPLE

### Electrolysis

***polarographic is defined as the electrolysis of a minute fraction of a solution around a small easily polarizable electrode.***

The electrolysis is confined to that part of the solution which is in constant with the surface of the small electrode, while the body of the solution remains unchanged.

The most familiar forms of electrolysis are those involved in electroplating bath or in the production of gases at electrodes. In all electrolysis, the chemical reactions which occur at the electrodes, are the result of flow of electrons from the external source of voltage.

In the galvanic cell or battery, however, chemical reactions in the cell produce a flow of electrons, which is current.

## LIMITING CURRENT

If one electrode is very small and other is large and the solution is not stirred. In such situation, the current will rise until the concentration of anions at the surface of small electrode approaches zero.

In such a quiescent solution, ions can only reach the electrode by diffusion. In addition, increase in applied potential will not increase the current linearly, since the current is now limited by the supply of anions. The magnitude of the current will depend upon the rate of diffusion of ions to the electrode surface. A plot of current-voltage curve in given in Fig. 26.1 and 26.2.

When the concentration of ions of the electrodes surface approaches zero, the concentration gradient across the interface approaches a constant. The value of this constant is a function of the concentration of ions of the solution.

Under such conditions, the small electrode is said to be concentration polarized i.e. the current flowing is limited by the concentration gradient of a specific ion at the electrode surface.

In the polarographic method the small polarized electrode is usually a mercury drop falling from a capillary. The capillary is selected to yield a new drop every 3-5 seconds. With each new drop a new electrode surface is provided, elimination all deposition products from the previous electrolysis.

Each drop is an exact replica of its predecessor, making the electrode characteristic reproducible.

## QUANTITATIVE USE.

The limiting current is actually the sum of three separate components:

The migration current, and

The diffusion current.

The residual current is the result of electrochemical reaction at the surface of the small electrode, independent of any specific ion reaction. The residual current is normally very small and is proportional to the applied EMF.

The migration current is the result of electrostatic forces which cause a flow of positive ions toward anode. The ions are said to migrate toward the opposing electrode. If a large excess of inert electrolyte is present in the sample, the electrical resistance between the electrodes is very small and current is carried entirely by the ions of the inert electrolyte.

However, since the electrons may reach to the electrode surface only by diffusion, the current which flows is very small and is actually dependent upon the rate of diffusion.

The diffusion current is the result of diffusion of the ions into the electrode-solution interface. The rate of diffusion is controlled by the concentration gradient between the ions at the surface of the electrode & those in the body of the solution.

Since the effect of the migration current is negligible and the residual current is constant at a specific applied EMF, the diffusion current alone may be considered. As indicated above, the diffusion current is controlled by a concentration gradient alone may be considered.

As indicated above, the diffusion current is controlled by the concentration gradient, which is in turn a function of the concentration of test ions.

Measurement of the diffusion current will yield a value directly proportional to the concentration in the body of the solution. Thus, the use of the polarograph as a quantitative instrument requires the application of a definite potential and the measurement of the resulting current. Fig. 26.3 shows a polarogram of copper solution at four different concentrations. Note that the wave height is directly proportional to the concentration of copper in each solution.

In quantitative analysis with polarograph, the current concentration relationship is determined on a series of standard of known concentration. A calibration curve is constructed & a proportionality factor is calculated, using the equation :

Where,

$C = K (D-D_0)$

C = Concentration

K = proportionality factor

D = diffusion current at C concentration

$D_0$ = diffusion current at O concentration

for subsequent determination of unknown samples the diffusion current is measure at the same applied EMF and under the same conditions & the concentration calculated from the proportionality.

## QUALITATIVE USE

The decomposition potential Ed, at which electrode reaction begins, is effected by the concentration of reducible ions. However the half wave potential (the potential at which the current is equal to one-half of its limiting value) is independent of concentration and specific for the particular ion being reduced. It may, therefore, be used for qualitative determination. $E^{1/2}$ can be determined graphically as shown in Fig. 26.3 and the polarogram of a mixture containing several ions is shown in Fig. 26.4.

## INSTRUMENTATION

### Circuity :

A circuit diagram is shown in Fig 26..5 the polarographic instrument has :

1. A source of EMF generally 0-3 volts.
2. A voltmeter for measuring the EMF.
3. A variable resistance for controlling the applied EMF.
4. A sensitive measurement galvanometer for measuring the resultant current.
5. A cell consisting of two special electrodes.

Most instrument also incorporate additional circuits for reversing polarity, selecting galvanometer range or automatically recording the current - voltage curves. Regardless of the complexity or cost of polarographic instrument have the five basic components as shown above.

## CELL

The electrode assembly normally used in polarographic work. The mercury level, which must be half constant, is controlled by the levelling bub. The cell contains the dropping mercury electrode and a reference electrode, generally saturated calomel, connected by means of a salt bridge.

## VARIABLES INFLUENCING THE DIFFUSION CURRENT

Relationship of cell design to current produced is influenced by following variables :

### 1. Size of the mercury drop

The amount of current flowing is proportional to the surface area of the drop. Therefore, drop size must be constant for quantitative work. Changes in capillary size or length or in static pressure on the mercury will influence the size of the drop.

### 2. Drop time

For reproducible result, the life of each drop should be 3-5 seconds. Shorter drop time will cause stirring action around the electrode, longer time will permit interference from vibration & adsorption of decomposition products. Drop time is affected by capillary characteristics, electrolyte composition & concentration, applied potential & pressure on the mercury.

### 3. Temperature

Changes in temperature affect the drop time and size of the mercury drop, the migration current, the rate of diffusion, and the adsorption of decomposition products Normally the cell and sample are immersed in a thermostated water bath to control temperature fluctuations.

Where the sample must be analysed without temperature control, a small thermometer is incorporated in the electrode assembly and suitable temperature correction calculated. Temperature coefficients have been found to be approximately 1.5% change in current for each 1°C change in temperature.

**4. Oxygen Interference**

For most polarographic determination, it is necessary to remove dissolved oxygen from the sample. Oxygen produced a very prominent polarographic wave over the region 0.1 to 1.8 volts VS. SCE.

Removal of the oxygen is accomplished by the addition of alkaline sodium sulphite solution or by bubbling an inert gas such as nitrogen through the sample.

**5. Reference electrode**

In order to accurately control the applied potential reference electrode must form one of the half cells in the electrode assembly. Generally a saturated calomel electrode is used. To prevent a shift in the half wave potential with SCE should be large & the salt bridge must have a low resistance junction.

## COMMERCIALLY AVAILABLE INSTRUMENTS

**1. Manual Instruments**

Requiring the manual control of the applied voltage through a rheostat. These are non recording instrument, so that polarograms are developed manually from galvanometer readings.

**2. Sargent Model III**

Wuses 1.5 V dry cells as a source of electromotive force (EMF). Current is indicated on a 300 mm curved scale. A moving galvanometer measures current over a wide range.

This instrument is versatile, yet has accuracy comparable to more expensive models. It is suited for routine laboratory analysis.

**3. Fisher Electrode**

A compact, self contained manual instrument. The cell and electrode assembly are mounted on the front of the cabinet, but may be dismounted for use in a water bath or special cell.

## RECORDING INSTRUMENT

These automatically apply an increasing potential at a constant rate. The movement of the chart recorder is synchronized with the voltage divider so that the current voltage curves are correctly plotted.

## OPERATING INSTRUCTIONS

Plug the main supply & dropping mercury electrode to their appropriate sockets. Switch on A.C. mains:

1. ***Polarographic cell***

   15 ml. sample degassed with nitrogen.

2. ***Indicating Electrode***

   Dropping mercury electrode suitable to give fine drops at the rate of 10-20 drops per minutes.

3. ***Reference Electrode***

   Saturated calomel electrode (SCE).

4. ***Speed of Chart run***

   1 inch in 100 sec. (i.e. 2 inch of chart/volt)

5. ***Current Sensitivity***

   0.5 to 1.0 μA for full scale deflection on the chart.

6. ***Range Multiplier : X1***

7. ***Damping : 3***

8. Variable polarizing potential range scanned it should be -0.3 to -0.6 volts.

Select direct position at Deriv/Null/Direct Switch. Switch on recorder motor and auto potentiometer motor, for polarising voltage is switched on as soon as the recording per reaches the desired starting voltage mark on the chart. Zero setting adjusted with the help of coarse & fine zero control knobs.

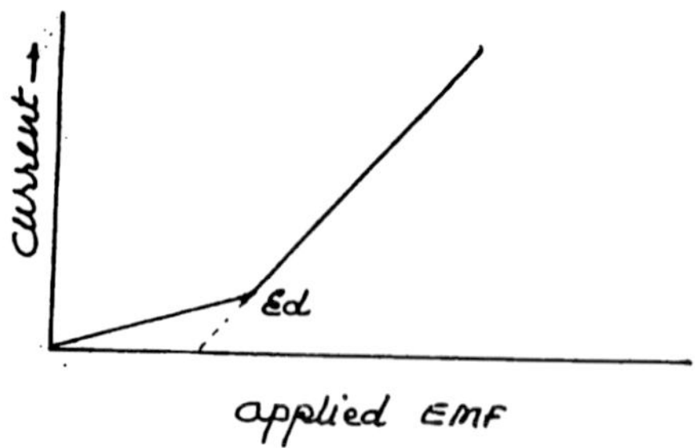


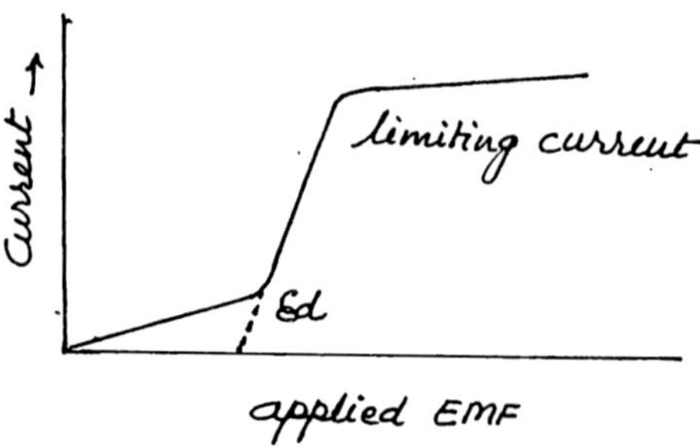


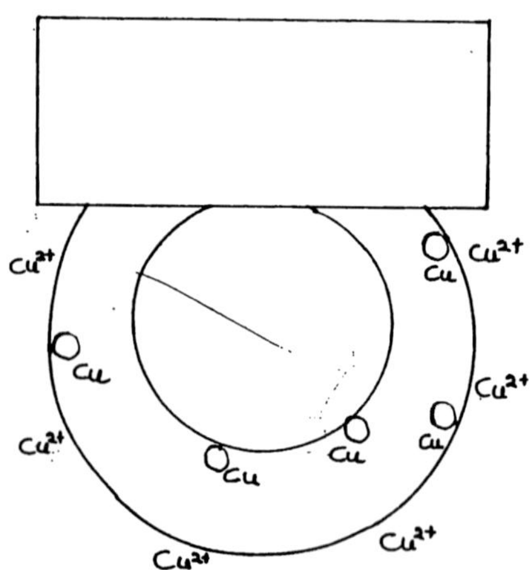


**Fig. 26.1 : Limiting Current**

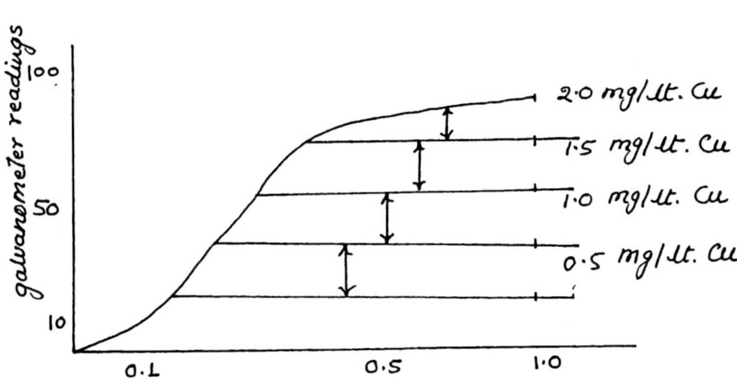


**Fig. 26.2 : Wavelength height at various concentrations**

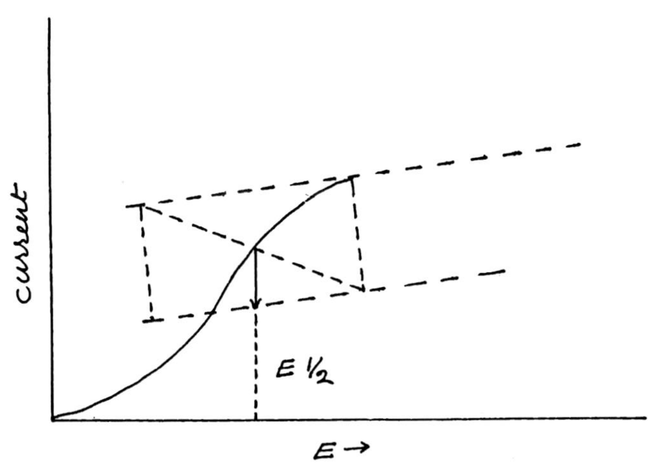


**Fig. 26.3 : Graphical determination of half rate potential**

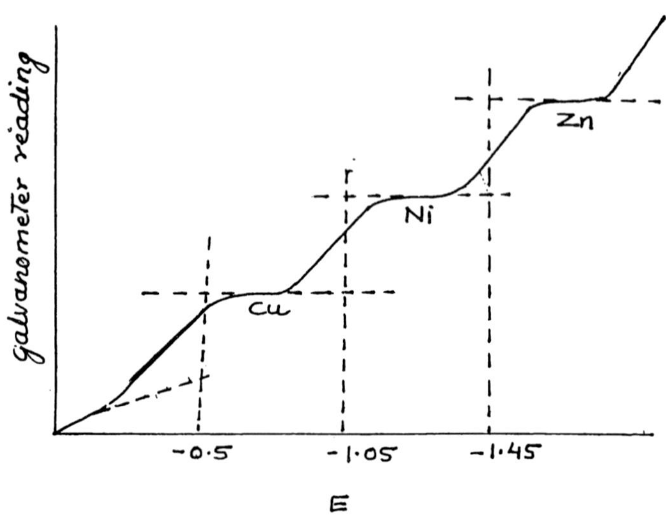


**Fig. 26.4 : Qualitative use of polarographic analysis**

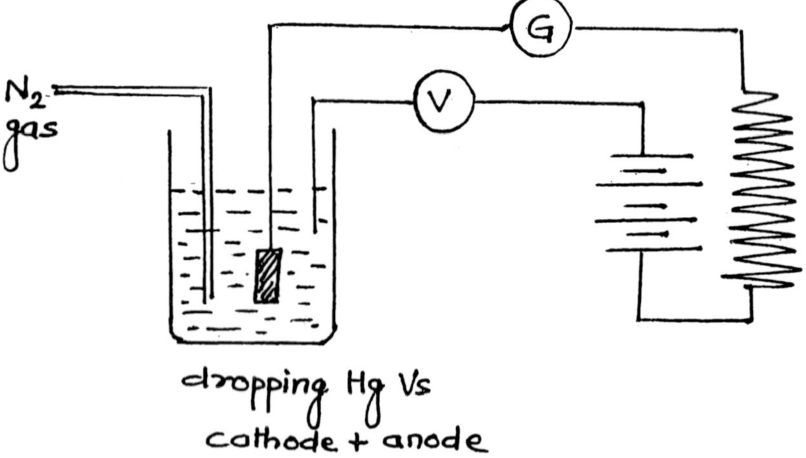


**Fig. 26.5 : Diagrammatic representation of polarographic analysis**

Chapter -27

# ANODIC STRIPPING VOLTAMETRY (ASV)

## INTRODUCTION

The detection of heavy metals in environmental samples has been long of intest to environmental scientists. To trace out the existance of metallic fraction in environmental sample at ppb and sub-ppb levels, a number of techniques exist.

Anodic stripping voltametry (ASV), colorimetry, neutron activity analysis, AAS are some of the techniques, which are being used with varying degree of success, convenience and cost.

Anodic stripping voltametry has an advantage over other techniques, as it does not require sample concentration, which otherwise involve possibilities of contamination from reagents and modification of the chemical state of species under investigation.

This techniques have importance due to :

1. Their firm, well defined physico chemical fundamentals in terms of kinetics and thermo-dynamics of electrode processes.
2. Its working mechanism, based on Faraday's law, provides high sensitivity, precision & methological accuracy in measurements.

ASV is a two stage process:

1. First often called the plating stage or electrode position, corresponds to the reduction of metal ions at a potential in the plateau region of a normal polarogram.

The electrode is generally a Hanging Mercury Drop (HMDE) or a mercury film on carbon substrate (MFE).

The metal gets concentrated in these electrodes as an amalgum and this pre-concentration gives ASV an increased sensitivity over polarography.

2. After a short rest period, the second or the stripping stage is began by varying the potential of the electrode, thus electro-oxidizing the metal into the ionic form, returning it to the solution.

Plating can be performed in two ways :

1. To plate or deposit all determinant in the sample into the electrode (total plating).
2. To plate a fraction of the determinant in the sample under reproducible conditions.

The later method is most common and is followed in most of the laboratories. For reproducibility, it is important to fix following plating conditions carefully-

1. Solution Volume.
2. Stripping Rate.
3. Mercury drop size or mercury film thickness of electrode.
4. Cell geometry.
5. Plating Time.
6. Plating Potential.
7. pH of the solution.

**PRINCIPLE**

The plating step is governed by the following first order kinetic equation :

$$Q_t = Q_o (1-e^{-Kt})$$

Where,

$Q_t$ = charge passed after plating time t

$Q_0$ = charge that would be passed with total plating and

$$K = \frac{DA}{\partial V_S}$$

Where,

D = diffusion coefficient of electroactive species

A = electrode area

$V_S$ = solution volume and,

$\partial$ = diffusion layer thickness.

$Q_t$ is related to $C_a$ the amulgum concentration of metal in following way :

$$C_a = \frac{Q_t}{nF.Va} = \frac{1}{nF.Va} \int_a^t idt$$

Where

Va = mercury volume.

n = number of electrons in electrode reaction.

F = faraday's constant,

i = plating current and

Q = bulk solution concentration.

The plating current 'i' is the analogue to the limiting current in a diffusion controlled process :

$$i = \frac{nF\ Cb\ AD}{S}$$

## TECHNIQUES

### 1. Cells

Normally a cell assembly for carrying out anodic stripping voltametry measurement consists of a working electrode, a reference electrode such as Ag/AgCl or Standard Calomel Electrode (SCE) and an auxiliary electrode which is most commonly a platinum wire.

In all cells, nitrogen inlet and outlet are provided so that nitrogen can be passed through the solution to remove oxygen before the experiment & passed over the sample during the experiment to prevent air re-entering the system.

The cell lid and the electrode holders are made such that the electrodes can be replaced in the same position if they have to be removed, usually they and the lid can be left in position and the cell/ bottom removed for cleaning, changing the sample etc.

Cell volumes between 50 ml. to 2 ml. are commonly used for ASV. Microcells taking 0.1 ml. solution have also been developed & used for analysis (Fig. 27.1).

## 2. Electronics

With the advent of low cost modern polarographic instrument, it has become essential to have a 'potentiostatic control' in the instrument so as to control the working electrode potential.

Modern instruments incorporate potentiostat which controls the potential at the working electrode solution resistance and enabling measurements in wider range of system.

In the potentiostatic electrode system (Fig. 27.2) the reference electrode of constant potential is placed near the working electrode & connected to an instrument through a circuit which draws no current from it.

Thus output of circuity within the instrument is then the voltage right a the tip of the reference electrode. The operational amplifier control loop applies sufficient compensation potentional to the counter electrode to ensure that the potential at the solution resistance is sufficiently high as to cause appreciable voltage drops when current flows through the solution.

## 3. Electrodes

1. ***Hanging Mercury Drop Electrode (HMDE)***

   The HMDE is being used to a considerable extent in ASV studies. HMDE's are made by collecting drops of mercury from a dropping mercury electrode and attaching them to a platinum wire embedded into a glass tube.

The electrode consists of a microsyringe with a micrometer screw for control of drop size. By turning the micrometer screw for control of a fixed number of divisions, reproducible drop sizes can be obtained.

2. ***Mercury film electrodes***

The development of Mercury Film Electrode (MFE) and its use in anodic stripping voltametry has changed the complextion of the technique to a considerable extend putting it on par with as complimentary to AAS.

The dial base material for a mercury film electrode is graphite. Among the graphite based electrodes, Glassy Carbon (GC) and Wax Impregnated Graphite (WIG) are widely used because of their non porous characters, which is a desirable property for forming a mercury film electrode.

**d. Supporting electrolytes**

Some of the common supporting electrolytes for ASV studies includes KCl, $KNO_3$, $CH_3COONa$, $H_2SO_4$, $HClO_4$, citrate buffer, NaCl, HCl, acetate buffer.

Purification of these is often necessary & is done by electrolysing the electrolytes using mercury pool cathode at a potential at which impurities are reduced.

## APPLICATION

### Water pollution studies

This method is used in determination of trace metals in various media like laboratory water, drinking water, natural water, sea water & sewage effluent. It can directly able to measure some of the trace metals.

ASV can be used to determine several elements in a single analysis. Mercury film electrodes are fairly satisfactory in such analysis. However, simultaneous analysis of trace metals raises a question of the formation of intermetallic compounds. These intermetallic compounds can ;

1. Change the stripping peak potential of the metal involved.
2. Lower peak current and,

3. Produce a new stripping peak due to the compound itself.

**Speciation Studies**

The term speciation used to denote the distribution of a metal among the different forms (species) present. Metal speciation can markedly affect the properly and behaviour of a metal in rivers e.g. toxicity, transport along a river, removal in the water treatments plant etc.

The likely forms in which the metals may be present are -

1. Metal ions complexed with other ions in water.
2. Metal ions adsorbed on to particles suspended in waters.
3. Those present in suspended solids.

Natural water contains a number of both organic and inorganic ligans e.g. humic acid, amino acid, vitamins, halides, carbonate, sulfate etc.

These legends play an important role by complexing the metals. The complexation can shift the stripping peak potentials. The shift depends on the ratio of the complexed to free metal concentrations.

Many complexes may be present in natural waters and potential shifts may be small. Therefore, complexes can in principle be determined by ASV.

Complexes may be arbitrarily divided into 2 classes :

1. Strongly - bound and,
2. Weakly - bound.

In practice, ASV can separately determine -

1. Free plus weakly bound metal and
2. Strongly bound metal.

Free and complex metals in rivers can also be determined by ASV.

**3. Air pollution studies :**

Metals also occur in aerosol as a result of biogeochemical cycle of metal in environment. It also present on dust particles, suspended and particulate matters of other type.

Therefore monitoring & physiochemical characterization of air-borne dust particles as a potential source of intoxication of man and mammalians has been of concerned to all those engaged in atmospheric pollution surveillance, for which, ASV is very much useful & fulfils most of the prerequisites. Table 27.1 shows some important metallic pollutants, can be determined by ASV.

**Table 27.1 : Trace metals determined in Water by ASV**

| *Metal* | *Status of Pollution* | *Method of Stripping* | *Type of Electrode used* |
|---|---|---|---|
| Pb | Pure lab water | DCS | RGCE |
| Fe | Pure lab water | DCS | RGCE |
| Pb, Cd, Zn, Cu | Pure lab water | DCS | RGCE |
| Cd, Mn, Ni, Zn, Sb, Tl, Cu, Pb, Sn, Bi | Pure lab water | PD | HMDE |
| Pb, Cd | Drinking Water | DC | MFE |
| Pb, Cd,, Zn, Cu | drinking + natural water | DC | HMDE |
| Pb, Cd, Zn, Cu | estuarine water | ACS | HMDE |
| Pb, Cd | natural water | PDS | MFE |
| Pb, Cd, Zn, Cu | ground, estuarine sewage & effluent | DCS | MFE |
| Pb, Cd, Zn, Cu | Sewage effluent | DCS | HMDE |
| Pb, Cd, ZN, Cu | sea water | DCS | HMDE |
| Pb, Cd, Zn, Cu, Bi | sea water | DCS | RGCE |
| Ni, Sn, U | sea water | DCS | HMDE |

HMDE - Hanging Mercury Drop Electrode

RGCE - Rotating Glassy Carbon Electrode

MFE - Mercury Film Electrode

DCS - Direct Current Stripping

ACS - Alternate Current Stripping

PDS - Pulse Differential Stripping

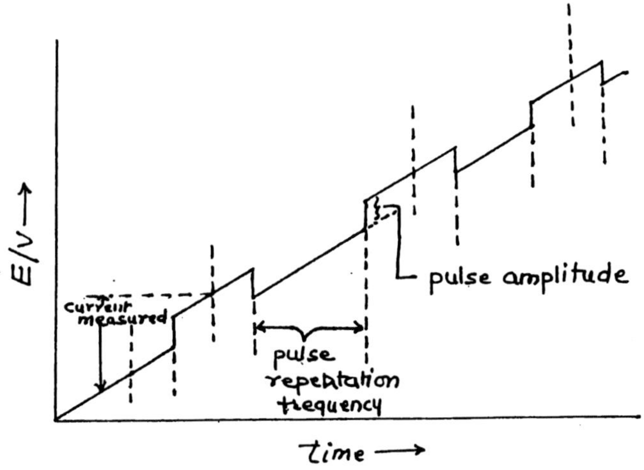


**Fig. 27.1 : Stripping potential waveform for DPASV**

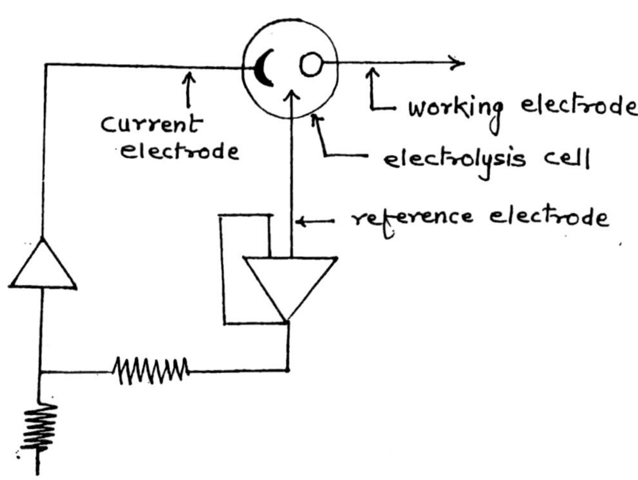


**Fig. 27.2 : Potentiostatic tree electrodes system**

## Chapter -28

# TOTAL ORGANIC CARBON ANALYSIS

### INTRODUCTION

Total carbon analyser is a major tool for the calculation of total carbon present in the water, wastewater & other effluents. It has been demonstrate by many investigators that the TOC correlates with the traditional 5 days BOD or 5 hrs. COD as an indicator of organic loading. In many waters & wastewaters investigators have successfully found the relationship (correlation) between TOC, BOD and COD especially where the wastes are consistent in composition.

This method is useful as one can quickly analyse the samples after getting them from the source.

### PRINCIPLE

The carbon present in the samples is converted into carbon dioxide by carrier gas i.e. oxygen which is measured by non-dispersive infra red analyser.

The homogeneous diluted samples when injected in the Total carbon channel or Inorganic Carbon channel, which are maintained at temperature of 950°C or 150°C respectively converts the carbon present in sample into carbon di oxide by carrier gas.

The percentage of carbon present in the samples are given in terms of percentage of carbon di oxide, which is measured by non-dispersive infrared analyser.

The difference of reading between total carbon & inorganic carbon gives organic carbon present in the sample.

## THEORY OF INSTRUMENT

### 1. Flow system

Two channels have different pressure regulators, pressure gauge, glow control valves & flow meters for carrier gas. The gas, then, flows through reaction chamber & then condenser to remove steam & then in turn in infrared analyser and in atmosphere.

### 2. Combustion tube

In the total carbon channels a high temperature furnace heats a ceramic combustion tube containing cobalt oxide deposited on an inert igneous substrate.

The temperature of furnaces is maintained at 950°C. The oxygen in the carrier gas oxides both organic & inorganic carbonaceous material to $CO_2$ & steam.

In the inorganic carbon channel, a low temperature furnace heats a glass tube containing quartz chips wetted with 85% phosphoric acid.

The temperature of furnace is sufficiently high for desired reaction but substantially below that required to oxidise in organic matter. The acid liberates $CO_2$ & steam from inorganic carbonates.

### 3. Electrical System

The power to both the furnaces are supplied through adjustable temperature controller assembly utilizing thermocouple as sensor.

The thermocouple & associated SCR & resistance constitute a bridge. Deviation of furnace temperature from the selected set point unbalances the bridge & causes the required increase or decrease in heater voltage.

### 4. Electronic circuit for Infrared Analyser

The infrared energy is produced by two separate energy sources. Then it is passed through optical filters to minimise interference & to the cells where one has non absorbing gas & other a continuous flowing samples & finally on detector.

The infrared energy absorbed being proportional to the $CO_2$ concentration. The detector is a gas microphone on "Luft principle".

It convert the difference in energy between sample & reference cells to a capacitance change. This capacitance change equivalent to component concentration, is amplified & indicated on meter or used to drive recorder.

**5. Recorder**

Output from infrared analyser is fed to the recorder which is sufficient to drive the pen motor for full span. The adjustment for chart speed & full scale span is done according to the requirement.

## PREPARATION OF WORKING STANDARD SOLUTION

**1. Organic Carbon Stock Solution :**

Dissolve 2.125 gms of analytical grade anhydrous potassium biphthalate ($KHC_8H_4O_4$) in $CO_2$ free water & dilute to 1 litre.

This stock solution contains 1000 mg/lt. (ppm) organic carbon stock. The stock solution of other concentration can be prepared by appropriate dilution.

**2. Inorganic Carbon Stock Solution**

Dissolve 4.404 gms. of analytical grade anhydrous sodium carbonate ($Na_2CO_3$) in 500 ml. $CO_2$ free water & add 3.497 gms. analytical grade anhydrous sodium bicarbonate to the flask & dilute to 1 lt. with $CO_2$ free water.

This solution contains 1000 mg/lt. (ppm) inorganic carbon stock, solution of other concentration can be prepared by appropriate dilution.

**5. Sample Preparation**

When the sample contains appreciable amount of suspended solids, it must be blended & 10 drops of conc. HCl is added for removing large concentration of inorganic carbon.

If the samples have large concentration of organic carbon, the sample may be diluted by $CO_2$ free water as sample should not contain total carbon more than 10,000 ppm.

**6. Operation**

Open shut off valve on oxygen cylinder & set cylinder regulator for output pressure of 10 psi, the pressure regulator of both channels have to set up at 3 to 5 psi & flow control valves at 100 to 150 ml./ min.

Place main power switch on & so the low temperature & high temperature furnace switch at ON. Turn 'ON' non-dispersive analyser & recorder. The set point of high temperature furnace is to be set at 950°C & low temperature at 150°C. Allow sufficient warm up time for stable drift free operation.

The instrument is them standardized for full scale span of 1000 ppm or 100 ppm for total carbon & 100 ppm or 10 ppm for inorganic carbon by injecting 20 microlitre of standard stock solution.

Turn sample selector valve to total carbon position. Rinse syringe with $CO_2$ free distilled water & then with sample.

Take 20 microlitre of sample in syringe & inject it in one stroke. Let the syringe be in that position until the peak for the injected sample registers.

Follow the same procedure for total carbon channel & inorganic carbon channel. The difference between the two registered peaks gives the total organic carbon present in it in terms of $CO_2$.

At least three successive injection should be done for one sample in each port to minimise the error.

# Chapter -29

# CARBON - HYDROGEN ANALYSER

## INTRODUCTION

The determination carbon & hydrogen in organic material depends simply on burning the sample in an oxidising atmosphere & measuring the combustion products carbon dioxide & water.

The first attempt to measure C & H was made by Gay-lussal & Thenord in 1810. A pelleted mixture of weighted amounts of sample & potassium chlorate ($KClO_3$) was dropped into a closed tube held in vertical position & heated at lower end using a spirit lamp, carbon dioxide, oxygen & water were collected over mercury & determined gasometrically.

Liebig in 1931 gave an improved method for the determination of carbon & hydrogen in organic material. Combustion was carried out in a 90 c.m. long hard glass tube in a flow of air (free from $CO_2$ & $H_2O$). Two third of the tube was filled with cupric oxide (CuO), maintained at 600°C by row of burners. The vapours of compounds were carried by air over cupric oxide.

Cupric oxide converted carbon into carbon di oxide & hydrogen into water. The product were passed over anhydrous calcium chloride which absorbed water & through solution of caustic potash (KOH), which absorb Carbon di oxide.

By weighing these tubes the amount of $CO_2$ of $H_2O$ were calculated & hence percentage of carbon & hydrogen in the original substance.

The attempts carried out by Liebig & Dennstedt were, not sufficient to determine the C & H in the organic material, because the apparatus was cumbersome & a single determination took several hours. To overcome their difficulties, as well as to enable small samples to be analysed, Pregl in 1910 his classic researches into the development of semi-micro methods for the analysis of carbon & hydrogen in organic compounds.

In Pregl's method, the sample contained in a platinum microboat was composed in a slow stream of oxygen (4 ml./min.). Oxidation was achieved by passing the gaseous mixture over the catalyst cupric oxide (CuO) & lead chromate ($PbCr_4$) maintaining temperature at 750°C by a furnace & a silver gauze maintained at 550°C.

The $CO_2$ & $H_2O$ in the issuing gas stream were collected in 'Pregl-type' absorption tubes & weighed. A 'conditioning' treatment which consists of burning off several samples of organic carbon materials, was necessary after replacing the tube before satisfactory results were obtained. The limitation of his method was that the chemical filling in the combustion tube was required fairly frequent replacement.

The next most notable advance in the method made by Belcher & Ingram who introduced the rapid 'empty tube' method. The filled oxidation tube was discarded & replaced by a specially constructed baffle tube, which ensures intimate mixing of oxygen with the organic vapours. The advantages of this technique were that the speed of determination was increased & apparatus required the minimum attention.

Apart from all these methods, the quantitative micro-analysis of the sample is determined by the recent advanced instrument known as 'carbon hydrogen analyser'.

The instrument can be used for the analysis of environmental organic pollutants for their carbon & hydrogen determination.

## PRINCIPLE

A definite amount of substance (10 mg) is weighted into a capsule crimped at both ends, placed in combustion tube & heated in stream of oxygen.

The resulting products of complete combustion, water & carbon dioxide, are absorbed in weighed tubes containing anhydrous magnesium chlorate & sodalime respectively & these are weighted again after absorption is complete, thus giving the weight of water & carbon di oxide produced, & hence the percentage of carbon & hydrogen in the original sample.

## APPARATUS

The apparatus for determination of carbon & hydrogen is presented diagrammatically as in Fig. 29.1 which essentially consists of three parts :

### 1. The scrubber tube

The oxygen used in the combustion is supplied by a oxygen cylinder fitted with a pressure reduction valve, pressure gauze & fine control knob.

In order to ensure the complete purity of the oxygen, it is first passed through a scrubber tube, 1/3 portion each of the tube is packed with Arcerite (Sodalime) and anhydrone (dehydrate) separated by a thin layer of cotton plug.

Traces of water & carbon di oxide from the incoming oxygen in the cylinder are absorbed in this tube & the purified oxygen is allowed to go in the system.

### 2. The absorption assembly

The absorption assembly is made up of three tubes known as ***absorption tubes***.

Absorption tubes are pyrex glass tubes provided with inlet & outlet & ground glass stoppers which allows inlet & outlet to be opened or closed.

First tube is filled with anhydrone (dehydrate) & packed with a small cotton plug at both ends. 1/3 portion of the second tube is filled with anhydrone (dehydrate) & 2/3 part is filled with ascarite (sodalime) separating each by a thin layer of cotton pad. Both ends are packed with a small cotton plug. 1/10 portion of the third tube is filled with anhydrone (dehydrate) & remaining portion with manganese dioxide ($M_nNO_2$) separated with a thin layer of cotton pad. Again, both ends are packed with cotton pad.

### 3. Combustion Tube

The combustion tube is made up of transparent quartz & held up in a vertical position with open and up. A roll of silver wire gauze having 1" wide & necessary diameter to fit the above of the combustion tube is inserted in end of the tube.

The tube is then filled with silver tungstate magnesium oxide upto the bottom of the enlarged. The quartz chips of 8-20 mesh are poured into the combustion tubes.

## OPERATION

The unit is connected to a oxygen supply cylinder. The furnace starts heating gradually after the electric power is switched on and attains the temperature of 900°C by gradually advancing the setting of the furnace control knob.

The absorption tubes are connected with the unit in order as shown in fig. The oxygen cylinder is opened in order to their relation to the oxygen flow. The operating button pushed to start the combustion cycle, after which the solenoid valve is opened with an audible click bound indicating the beginning of the sweep cycle & oxygen flow through the system.

The oxygen flow rate is adjusted to 100 ml./min. After three minutes a buzzer sound indicated the end of operation, closing the selenoid valve, stop of oxygen flow & the end of the operation. Thus, ensuring the completion of sweep cycle.

The reset switch is depressed to stop the buzzer. The same is repeated 2-3 times to bring the whole system in indication before the sample run.

After the initial conditioning absorption tubes are deteched for initial weighing & replaced.

The cap is replaced immediately & locked. As soon a as the operating button is pushed, an intense flash of light is emitted indicating rapidly burning of the sample inside the combustion tube.

After completion of sweep cycle, the stoppers of the absorption tubes are closed & reweighted.

The increase in weight indicate the presence of carbon dioxide & water.

The percentage of carbon & hydrogen is calculated as follows:

$$C\ (\%) = \frac{\text{increase in weight of ascarite tube}}{\text{weight of sample}} \times 100 \times 0.2729$$

$$H\ (\%) = \frac{\text{increase in weight of dehydrate tube}}{\text{weight of sample}} \times 100 \times 0.119$$

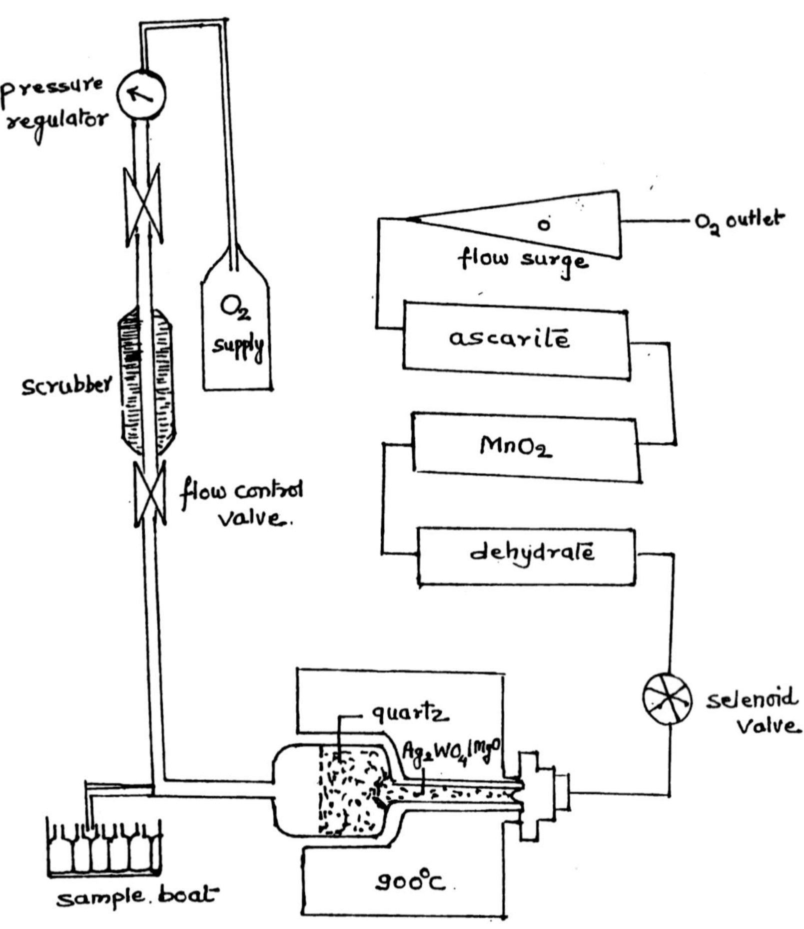


**Fig. 29.1 : Schematic flow chart for TOC analysis**

## Chapter -30

# X-RAY FLORESCENCE (XRF)

## INTRODUCTION

As the degree of industrialisation increases, contamination of atmospheres becomes more & more severe, until it now effects the living organics in number of undesirable ways.

Suspended particulate matter in air is recognized as a major pollutants which can have pronounced effect on health & environment. Toxic trace elements such as V, As, Se, Cd, Sb, Hg & Pb etc. are well known to be hazardous to human health when present in air & water.

The heavy metals, such as Pb & Cd are toxic not only to aquatic life & fish but also to the human body. With increasing concern about potential health hazards caused by foreign substances in air & water, the role of X-Ray Fluorescence (XRF) spectroscopy has become increasingly important and has come into extensive use for accurately monitoring the concentrations of trace elements in the environment.

## ANALYTICAL METHOD

For XRF, the sample can be collected on filter paper for particulate pollutants. Data analysis for XRF is less complicated. Use of XRF in air pollutants analysis is greatly increasing due to following reasons :

1. The direct analysis of filter deposits with no need for sample preparation.
2. The non-destructive analysis, which permits sample to be retained for further analysis & for future reference.

3. The fairly uniform detectability across the periodic table with the ability to analyse all elements.
4. Precise, accurate, fast & selective analysis.
5. The availability of commercial instruments that permit the analysis of sample for a large number of trace elements in relative short intervals (100 scc.) at the low cost.

## EXPERIMENTAL TECHNIQUE

In XRF analysis, a sample is put in a beam of X-ray energetic enough to knock an electron out of one of the inner electron shells of the elements in a sample.

The hole left by displaced electron is usually filled by an electron from an outer electron shell to high energy in the same atom.

When this happens an X-ray photon or radiation is emitted (fluorescence) whose energy ($h\theta/\lambda$) or wavelength ($\lambda$) is characteristics of the elements involved (i.e. $1/z^2$, where z is the atom number of the element).

## PRINCIPLE OF ANALYSIS

XRF analysis is based on the measurement of wavelength (or energy) & intensity of characteristic X-Ray emitted by a sample which has been excited by photons or charged particles such as electron proton & particles (proton includes x-ray emission is called as PIXE techniques) A description of photon techniques can be naturally divided into three areas:-

1. X-ray generation (high voltage) sealed off-X-ray tube, radioisotope source of X-ray tube-fluorescence combination.
2. Wavelength (or energy) & intensity, number of detected X-rays) measurements.
3. Data interpretation.

Two available measurement techniques are wavelength dispersion or energy dispersion.

Energy E in KeV is euqal to -

$$E = \frac{12.398}{\lambda \text{ in } \mathring{A}}$$

Where,

Å = $10^{-8}$ c.m. &,

λ = Wavelength

**a) Wavelength dispersion**

In wavelength dispersion, each characteristic line emitted by the atom present in the unknown sample is diffracted at definite angle by certain family of planes containing atoms, having suitable interplaner spacing of the proper analysing crystals.

By selecting suitable crystals, it is possible to make measurement over wavelength range of about 0.25 A to 150A. The resolution of crystal spectrophotometer is more than adequate to distinguish each element without interference from either α or β lines of neighbouring elements.

When spectrophotometer is operated in air, the spectral region conveniently had wavelength range of 0.4 to 2.5 Å.

Element having 21 < Z (atomic number) > 55 can be detected by their k spectra while those having Z > 55 can be detected by their L spectra. Element having Z < 21 & upto Z = 9 emitting characteristic wavelength λ greater than 2.5 A, requires helium flushing or evacuation of the spectrophotometer to avoid significant absorption in the air paths.

**b) Energy dispersion**

Energy dispersion identifies the elements present in sample directly by their characteristic photon energy. This is done with an energy sensitive detector which emits a pulse each time and X-ray photon is detected and the amplitude of the pulse is proportional to photon energy.

Virtually all the energy dispersion measurements being made today use the solid state detectors. Furthermore, because the amplified pulse height produced are proportional or X-ray energy (or wavelength), it is possible to transmit separately pulses of different height triggered by x-rays of different energies (or wavelenghtly using a single channel pulse height analyser (PHA). Multichannel analyser (MCA) includes number of PHA.

Pulses correspond to each λ (or E) from PHA or MCA next pass into decade scaler which counts one output pulse for ten number of input pulses. The counts accumulated in the scaler, at the end of run of fixed time (adjusted by the timer) counting, represents the desired information of x-ray intensities of that particular λ (or E), which is proportional to the concentration of that element present in the sample.

The digital counts on decade counting tubes can be directly read. The printed output is especially advantageous when a great number of measurements are taken on a routine analysis.

The printed read-out is accomplished by use of teletype printer, which transfers the numerical information in decade scalers into the printer.

Auxiliary information for identifying purposes i.e. crystal angle, name of the elements etc. can be printed alongside the accumulated number of results. For continuous presentation of spectrum, the instantaneous average counting rates (i.e. x-ray intensities) are converted into continuous electric current & curve of λ (or E) versus intensity is plotted on an electrical recorder.

## PREPARATION OF CALIBRATION STANDARDS FOR AIR SAMPLES

The X-ray analysis of air pollution particulates on filter substrates, the reagent blank' consists of only the impurities in the filter material.

It is desirable to collect the sample on the purest filter available. It is noted that the membrane filter papers are of considerably higher purity as compared to that of glass fibre filter paper. A measured volume of gravimetrically prepared solution standard using pure material is micropipetted by the multidrop technique onto a membrane filter so as to homogeneously impregnated the filter.

This is rapidly dried under controlled condition yielding a single element or multielement calibration standards of high purity accuracy & homogenecity of desired concentration in $\mu g/cm^2$.

Standard can be supplied encapsulated between stretched x-ray transparent films & mounted in sample holders compatible with the user's XRF spectrometer. This film formed from vacuum evaporated elements & compounds prepared on mylar substance are also used with XRF, PIXE & $\alpha$ particles induced XRF techniques.

Routine collection of aerosol particles on a membrane type filter is a straight forward problems requiring that a known volume of air be drawn through the filter using a vacuum pump at a flow rate of 5-10 litres/minutes/c.$m^2$ for about 2-6 hours.

## XRF SPECTROMETERS & ITS APPLICATIONS

### 1. Environmental Protection Agency (EPA) Multichannel Wavelength XRF spectrometer

The system consists of an array of fixed wavelength spectrometers optimized for simultaneous analysis of 16 preselected elements & a computer operated scanning channel for determination of any number of additional elements.

A leading device permits automatic handling of batches of upto 100 frame mounted 47 mm filter samples. Instruments operation, data processing, correcting for the background & interference effects & converting X-ray intensity to concentration & printed out of results are controlled by a minicomputer.

The system permits rapid, direct, non-destructive, precise, accurate, fast & selective multi-element analysis at high spectral resolution, a significant advantage with air pollution samples which typically contain several dozen elements at a wide range of concentrations.

It simultaneously analyse (in 100 seconds counting times) thirty elements of greatest environmental interest, with very little sample preparation, in the range of concentration of 2 to 30 ng/$cm^2$ deposited on the filter paper.

**2. FDXRF System**

Ortec Corporation Ltd., Tennessee, USA manufacture the model 6110 Tube excited fluorescence analyzer (TEFA). The major components in it are x-ray tube, Si (Li) detector, specimens, system control, liquid Nitrogen Dewar flask & electronic counting system including amplifier, multichannel analyser connected to :-

1. X-Y plotter.
2. Cathode ray tube display &,
3. Page printer through computer.

Because of its energy dispersive detector system the TEFA simultaneously quantitatively analyses 82 elements @ 1/min, sodium & heavier plus trans uranium elements over a broad range of concentration from parts per million (ppm) to 100%.

## BROAD AREAS OF APPLICATION

Not only in pollution of air or water, but we are becoming more & more concerned with the harmful trace elements in agricultural & processed food as well as in marine seafoods.

XRF techniques also provided an ideal analysis system for samples of food, agricultural, products, fertilizers, solid wastes, soil, plant, forensic & medical science, pharmaceuticals & petrochemicals & semi conductor physics.

The analytical capability of the techniques has been demonstrated with a large variety of sample types from :

1. Mobile pollution sources such as diesel engines aircraft.
2. Stationary sources including incinerators, power plants, chemical process plants, cement dust and,
3. Ambient air sampling at different location.

XRF may be used to identify the contributions from the different sources for the observed concentrations of a particular toxic elements.

One can also obtain information about the mechanics of air diffusion, turbulence, vortex motion etc. & their dependence on geophysical & meteorological conditions.

## SCOPE

Because of its potential utility, this technique is finding a place as a gem technique in global environmental monitoring (GEM) programme.

The instrumentation is the most expensive but cost per elements for quantitative analysis are the lowest. The great advance in capabilities have not come from chemistry but from electronic (solid state detector, pulsed x-ray tube, integrated circuity, high speed computer).

As XRF instrumentation and quantitative techniques would progress, it would be possible to analyse more of low $Z < (9)$ elements.

X-ray lasers have been the recent subject of speculation & may be applied to this field in the next decade for improved performance. Perhaps it is there very developments, which make the field so fascinating & make contemptation of what is over the horizon in the distant future so exciting.

# Chapter -31

# GEIGER MULLER COUNTER

## EXPERIMENTS

1. Determination of dead time of GM counter.
2. Determination of back scattering co-efficient of a material for β-radiation.
3. Verification of inverse square law for λ-radiation.
4. Determination of self absorption coefficient of a β - source.
5. Determination of half life of a radioactive source.
6. Determination of mass absorption coefficient of aluminium for λ & β - radiations.
7. Determination of end point energy of β - radiation by finding out the range in aluminium.

## COMPONENTS OF THE UNIT

1. Main unit- GM Counting supply unit (Fig. 31.1 and 31.2).
2. G.M. Counter.
3. Detector stand M 1900 A holds the end window GM counter, source & aluminium absorber.
4. Lead castle C 1700 A provides scheduling around the source of GM counter.
5. β - reference standards R 2100 (set of 5-β-emitting isotopes).
6. Gamma reference standards R 23008/ (set of 66 λ-emitting isotopes).

7. Aluminium absorber A-1750 (set of 14-aluminium absorbers).
8. Stainless steel planchets P -1850 for depositing sample.
9. Stainless steel pans for depositing samples.

## PRECAUTIONS

1. The connection to anode & cathode must be reliable (free from variable contact resistance and microphony) and adequately insulated.
2. The GM counter should be surrounded by at least 4 centimetres of lead line with 1 millimetre of aluminium in order to obtain consistent measure of the background counter in any particular environment.
3. All radioactive substances should be removed from the immediate vicinity.
4. For accurate quantitative work at moderately high counting rates, the paralysis time to the counting equipment must be greater than the resolving time for GM counter.

## CONSTRUCTION & PRINCIPLE

The basic unit of the counter is made up of a cylinder of a silvered glass or brass cylinder with a wire in its centre.

The central wire serves as anode & the cylinder serve as a cathode. The two electrodes are enclosed in a glass tight envelope typically filled to a pressure of 80 mm of Argon gas plus 20 mm of methane or alcohol or 10% of chlorine.

At the base of the cylinder of mica end window about 2.5 c.m. in diameter and 2-3 mg/c.m. in thickness. This serve as the entry point for the radiation. When a potential of 800-2000 V is applied to the central wire anode, the ionising particles enter the active volume of the GM counter, where the colloid with the filling gas and produce ion pairs. These particles, then migrate toward the appropriate electrode under a voltage gradient.

The mobility of the electrons is quite high and under the influence of a high voltage gradient, it soon requires sufficient velocity to produce new pair of irons upon collision with the argon atoms.

Original ionising particles entering the active volume of the counter gives rise to an avalanche of electrons travelling towards the central anode. Photons are emitted when the electron strike the anode, due to which the ionisation is spread throughout the tube.

These processes produce a continuous discharge which fills the whole active volume of the counter in less than a microsecond. Each discharge build up to a constant pulse a maximum amplitude 10 V to 80-100/U sec. duration.

These pulses are displayed on the digital scale or a meter without any amplification.

## DEAD TIME OF THE COUNTER

The movement of the positive ions (i.e. protons) due to this heaviness is much slower (200 μ sec. than the negative ions (electorns) (50-100 μ sec).

During the time of electron avalaunche, it is collected on the anode, the positive ions move only a short distance on their way to the cathode. But during their travel time, the positive ions form a virtual sheat around the anode, which effectively lowers the potential gradient to a point where the counter becomes insensitive to the entry of further ionizing particles. This is called the ***dead time of the counter***.

Counting rates are limited to about 15000 counts/mm. because of the long dead time of 200-270 μ sec. There is a large reduction in true count rate as the maximum count rate is approached.

The GM counter is more sensitive to β - particles, but for α - particles, it much less than that of the scintillation counter.

## OPERATIONAL CONTROLS

### Controls on front side

i) ***Mains*** -

When the switch is depressed, the unit get 'ON'.

***EHT volts***

This is indeed, Electric high Tension volts. There are two switches:

a) ***Coarse***

This is a rotary control to adjust EHT output in steps of 150 V each. The EHT output ranges from 0-1500 V. It has OFF position on the left hand side.

b) ***Fine***

It controls EHT output in steps of 15 V (range from 0-150 V).

***Count ON/OFF***

This is a side way moving switch. It permits counting on the meter when switched to 'ON' position & stop counting in 'OFF' position.

***Reset***

This is a process switch & when pressed, the displayed counts return to zero.

***Paralysis time***

This is a rotary control & provides 3 stages of paralysis time. There stages are -

a) 250 microseconds.

b) 350 microseconds and,

c) 560 microseconds.

For use with various types of GM counter, it has an OFF position, which is operative only when counts are recorded with counter other than GM input.

**Controls of rear side**

***EHT out***

This is coaxial connector and provides direct EHT to a GM detector, when the load resistor is provided externally.

This connector is used only when the GM counter requires a load register other than 4.7 mega ohms.

***Input***

This is a receiving coaxial connector (socket) it receives input signals from sources other than GM inputs.

*EHT check*

This is a banana socket and is used for checking the EHT directly

The above three controls are used only under special conditions

*GM Socket*

This is a receiving coaxial connector (socket) for the GM detector. It receives the detected pulses from the GM detector. This connector provides the EHT to the detector through a built in 4.7 mega ohm load resistor.

## OPERATIONAL PROCEDURE

1. Put all the switches on 'OFF' position in the beginning.
2. Connect the GM detector to GM socket on the rear side.
3. Connect the main chord of the (GM system) unit with the mains (power supply of 220 - 240 V) and switch 'ON'
4. Place the radioactive substance under the counter in holding tray.
4. Depress the switch 'MAINS' to put on the unit.
5. Wait for 15 minutes till the unit is stabilized.
6. Select the desired paralysis tune (according to GM detector) increased the applied voltage till input sec. begun to register. Note this voltage (the starting point).
7. Press & release the RESET switch.
8. Switch on the count ON/OFF switch to ON position. This voltage is the starting potential.

   Simultaneously start the STOP Switch, when impulses being to register. If the counts are recorded more than 25000 mm, lower the voltage and remove the sample to avoid damage to the counter.
9. When the desired counting time is over, switch off the count & stopwatch simultaneously.

## IMPORTANT TO NOTE THAT

1. If the counting is to be made through some other detector not the GM detector, the detector is to be connected to 'input' socket (not to GM) on the rear side the EHT paralysis time switches are kept in OFF position.

2. ***EHT out***

   This connector is used only when such power GM tubes are used, which require load register other than 4.7 mega ohm. The detector is connected to input socket on the rear side.

## PREVENTIVE MEASURES & MAINTENANCE

1. Keep the unit free from dust & moisture.
2. Blow the instrument with clean compressed air association.
3. Clean the EHT connector with absolute alcohol before use.
4. Check the firmness of the switches. They should not be loose.
5. Check the cables occasionally for damaged parts.

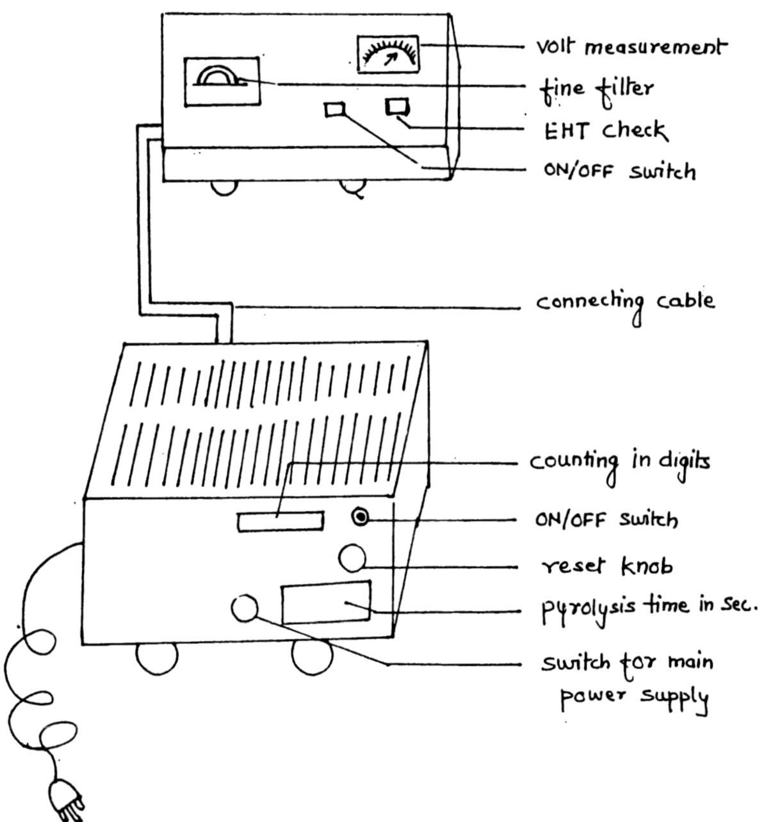


**Fig. 31.1 : G.M. Counting Unit**

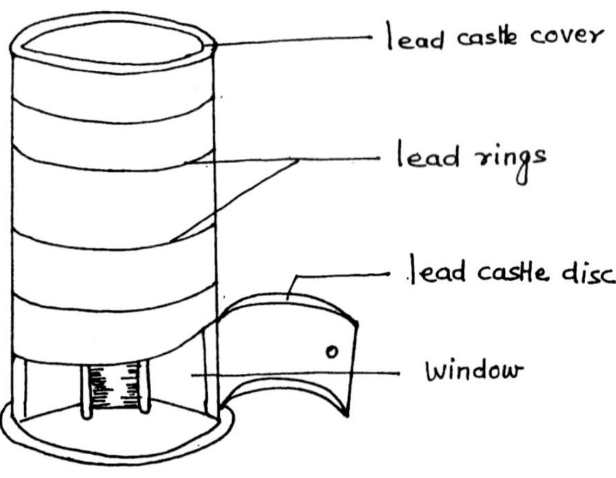


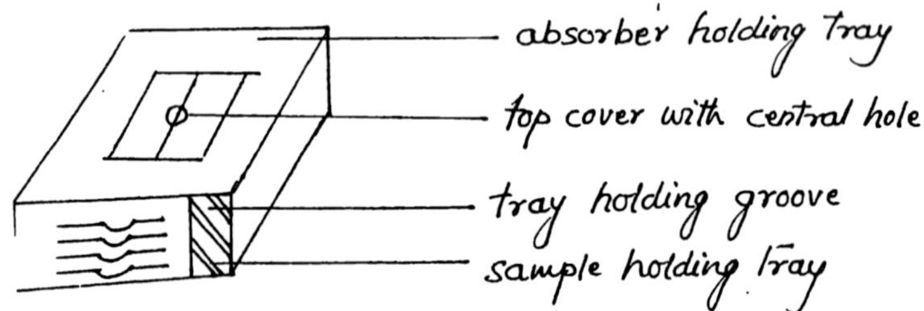


**Fig. 31.2 : G.M. Counter**

# Chapter -32

# MICROTOMY

## OPERATING INSTRUCTIONS

The microtomy depends on the nature of the knife edge, the till of the knife, the temperature, material section & embedding.

## PREPARATION OF MATERIAL

1. Sharpen the knife on the glass stone by using the clove oil.
2. Trim the block & fix it on the block holder.

## PREPARATION FOR MIRCOTOME

1. Adjust the desired thickness of section by turning the concerned knob and read the scale on right hand top (between the block holder & wheel) where -

   1 Division = 1 micron

2. Bring the object carrier back by rotating the reversing handle clockwise upto the limit & pull the level up.
3. Mount the block holder and object by adjusting the screw of the block holder.
4. Insert the knife laterally into the knife holder and adjust the inclination of the knife with respect to object by raising or lowering the thumb screw.
5. The knife must be titled so that there is clearance between the cutting facet next to the block of the material and the surface.

   Skipping of a section or the cutting of alternate thick or thin section is usually due to insufficient or too excessive tilt.

6. Tighten all screws, holding the specimen block the knife the knife holder before cutting sections.

   The angle below the blade is the clearance angle and the angle above it is the rack angle. The clearance angle ranges between 30° to 80°. For hard materials, the rack angle of 22°-25° are used & for soft materials, rack angle may be increased to 45°.

7. Rotate the wheel clockwise with the right hand for cutting the sections. The gap between the object and knife edge can be adjusted by turning or-reversing the handle. There is a wheel lock for locking the crank wheel to prevent the damage to the material.

   If the feed block has come nearer to the pin, stop the section cutting & drive the feed block away from pin by turning the reversing handle clockwise & then bring the razor nearer to the block by shifting the knife holder.

## Chapter -33

# MEASUREMENT OF TEMPERATURE IN ENVIRONMENTAL SAMPLE

### INTRODUCTION

Temperature plays an important role in biological activities, fish life, limnological studies etc. Temperature governs solubility of oxygen, carbon dioxide, bicarbonate, carbonate, carbonate equilibrium.

### APPARATUS

A thermometer having a quick response and having atleast divisions for 0.5°C.

### MEASUREMENT PROCEDURE

#### Air temperature or ambient temperature

Hold a dry thermometer at least one meter above ground, keeping it away from any heat source or direct solar radiation & read.

#### Water temperature

Thermometer is dipped directly into water body & a constant reading is noted. Temperature can be measured in the sampling bottle immediately after taking the sample. Ensure that container attains the temperature of the water sample.

### EXPRESSION OF RESULTS

Temperature may be recorded to the nearest of 0.5°C.

## PRECAUTIONS

1. While taking temperature of water sample, ensure that the mercury bulb of the thermometer is dipped in the water.
2. Before making any temperature measurement, wipe the thermometer with tissue paper or filter paper.
3. Whenever the ambient air & water temperature both are to be measured, it is suggested that the earlier should be followed by later.
4. Do not touch mercury bulb, while taking the readings.

## Chapter -34

# MEASUREMENT OF HYDROGEN ION CONCENTRATION OR pH IN THE ENVIRONMENTAL SAMPLES

### GENERAL

pH is the negative logarithm of hydrogen ion concentration (activity of hydrogen ions) in moles per litre at a given temperature. The pH of natural water is controlled by the carbon dioxide/bicarbonate equilibria & usually ranges from 4.0 to 9.0. The majority of waters are slightly basic due to presence of bicarbonate & carbonates. pH affects the chemical & biological properties of liquids & hence is an important estimation e.g. for controlling corrosion, water & waste treatment processes.

Electrometric & colorimetric methods are available for the measurement of pH.

### ELECTROMETRIC METHODS

*Principle*

The pH is determined by measurement of the electromotive force of a cell comprising an indicator electrode, an electrode responsive to hydrogen ions (glass electrode) immersed in the test solution & a reference (usually calomel) electrode. Contact between the test solution & the reference solution is achieved by means of a liquid junction, in the reference electrode. A difference of 1 pH unit produces a potential charge of 58.16 mv at 25°C. The electramotive force is measured with a pH tro read directly as pH value.

### *Interferences*

The glass electrode is free from interferences such as colour, turbidity, colloidal matter, oxidant reductants. In case of salinity is high a sodium ion error at pH 10 & above has to be compensated.

Temperature affects the potential at electrodes & the ionizaiton in the sample. The first effect can be overcome by a temperature compensation adjustment provided on the most of the instruments. The second effect is inherent in the sample & is taken into consideration by recording both temperature & pH of each sample.

## APPARATUS

1. ***pH meter***

   Mains or batter operated model (for field measurement) with temperature compensation adjustment & readily to an accuracy of 0.1 pH units.

2. ***Glass electrodes***

   This must be compatible with the pH meter used & should cover the entire pH range with minimum sodium ion error.

3. ***Reference electrode***

   Calomel electrode is widely used. Less concentrated solution of potassium chloride e.g. 3.5 M KCl or 350 g/l. are more satisfactory than the saturated solution usually used which very often leads in clogging of the electrode.

## REAGENTS

1. ***Saturated pH buffer solution***

   This can be prepared from buffer tablets which are commercially available. However, an analyst can prepare his own solution referring to standard texts.

## PROCEDURE

1. Rinse the electrodes with distilled water & dry be gentle wiping with a soft tissue paper.
2. Standardized the instrument with the electodues immersed in a buffer solution with a pH close to that of the water to be tested.
3. If manual temperature compensation is provided set it to the temperature of the buffer.
4. Check the electrode response occasionally by measuring a second buffer solution of different pH.
5. Rince with distilled water & dry with tissue paper after every measurement. The difference in reading for the standard buffers should not exceed 0.1
6. Immerse the electrode in the sample
7. Set the temperature control adjustment knob to the temperature of the sample. Taking final reading after 30 sec.

## EXPRESSION OF RESULTS

The pH value should be reported to the nearese 0.1 pH unit. The temperature of the sample at the time of pH measurement should be indicated.

## PRECISION & ACCURACY

The precision & accuracy attainable with a pH meter depends on the type of condition of instruments. A precision of ± 0.02 pH unit & an accuracy of ± 0.05 pH unit can be achieved. However ± 0.1 is satisfactory. Hence the pH should be reported to the nearest of 0.1 pH is satisfactory. Hence the pH should be reported to the nearest of 0.1 pH unit.

## PRECAUTIONS

1. Before making any measurement rinse the electrodes with distilled water & dry it with the tissue paper.
2. Ensure adequate level of filling solution in the reference electrode.
3. Always keep the electrode in distilled water, never allow them to dry out.

4. Standardize the instrument in case of any doubt with the pH buffer closer to the pH of the test sample.
5. Adjust the temperature compensation knob before reading.
6. After immersing the electrode, wait for sometime so as to stabilize the system. Take the reading as soon as it becomes constant.
7. As far as possible it should be measured at the site of sampling.
8. No chemical preservation of the sample is permissible.
9. Universal glass electrodes ensure good results.

Chapter -35

# MEASUREMENT OF MAGNESIUM & CALCIUM IN ENVIRONMENTAL SAMPLES

## MEASUREMENT OF MAGNESIUM

### General

Magnesium is a common constituent of natural water. It is one of the important contributors to the harndess of water. Magnesium salts break down on heating to form scales in boilers. Concentrations greater than 125 mg/lt. can also exert a cathartic & diuratic action. The magnesium concentration in water way vary from zero to several hundred milligrams per litre, depending on the source of the water.

Chemical softening of ion exchange reduces the magnesium & associated hardness to acceptable levels.

### Magnesium by Calculation

For routine estimation of magnesium concentration in water, the magnesium is calculated from the EDTA calcium & total hardness titration methods.

Magnesium = Total hardness - Caclium hardness, Expressed as Mg/l. $CaCO_3$

## MEASUREMENT OF CALCIUM

### General

The presence of calcium in surface water results from passage through or over deposits of lime stone, dolomite, gypsum & gypsiferous shale. The concentration of calcium may range from zero to several hundred milligrams per slittre depending on the source. Small concentrations of calcium carbonated prevents corrosion of metallic pipes by laying down a protective coating. On the other hand, appreciable quantity of calcium salts break down on heating to form harmful scale in boilers, pipes & cooking utensils. Calcium contributes to the total hardness of water.

The atomic absorption spectrophotmeter method is the most accurate method available for the calcium determination. But the EDTA & permagnate titration method also give good results for control & routine applications. Due to simplicity & rapidity of other EDTA method, it has become first choice of the analyst.

### PRINCIPLE

When EDTA (Ethylene diamine tetra acetic-acid) is added to water containing both calcium & magnesium, it combine first with the calcium. By titrating with EDTA, calcium can be determined directly. When the pH is sufficiently high so that the magnesium is precipitated as the hydroxide & an indicator is used which combines only with the calcium. Indicator shows a colour change when all of the calcium is complexed by EDTA at a pH of 12 to 13.

### INTERFERENCE

Under conditions of test, the following concentrations (Table 35.1) of the ion cause no interference with the calcium hardness determination :

| *Metallic ions* | *mg/lt.* |
|---|---|
| *1* | 2 |
| Copper | 2 |
| Ferrous iron | 20 |
| Ferric iron | 20 |

| *1* | 2 |
|---|---|
| Manganese | 10 |
| Zinc | 5 |
| Lead | 5 |
| Aluminium | 5 |
| Tin | 5 |

## REAGENTS

### Standard EDTA titrant (0.01 M)

Dissolve 3.723 g analytical reagent grade EDTA (sodium salt) ($Na_2H_2C_{10}H_{12}O_8N_2.2H_2O$) in distilled water & dilute to 1 lt. Standardize against standard calcium solution. This solution is approximately 1.00 mg $CaCO_3$/ml.

Orthoposphate will precipitate calcium at the pH of the test. Strontium & barium also interfere Alkalinity in excess of 30 mg/lt. may cause an indistinct & point with hard waters.

### Standard calcium solution (0.01 M)

Weigh 1.000 g anhydrous calcium carbonate $CaCO_3$. Powder into a 500 ml. erlenmeyer flask place a funnel on the flask & add slowly HCl untill al the $CaCO_3$ has dissolved. Add 200 ml. distilled water & boil for few minutes to expel out $CO_2$. Cool & add few drops of methyl red indicator & adjust to the intermediate orange colour by adding 3 N $NH_4OH$ or 1+1 HCl as required. Make up the volume upto 1 lt. in a volumetric flask. This standard solution is equivalent to 1.00 mg $CaCO_3$/1.00 ml.

## INDICATOR

Mureoxide (ammonium purpurate) was the first indicator available for the detection of the calcium end point, but due to difficulty in recognizing the end point, it is rarely used these days. Eriochrome Blue Black R or Solochrome Dark Blue (also called Calcon) gives an improve end point, changing colour from red to pure blue.

## ERICHROME BLUE BLACK

Prepare a stable of the indicator by grinding together in a mortar with 200 mg powdered dye & 100 g solid sodium chloride to a 40-50 mash. Store in a tightly stoppered bottle. Use about 0.2 to 0.4 g of the mixture of each titration.

During the course of titration the colour changes from red through purple to bluish purple to a pure blue without any trace of reddish or purple tint. The pH or some samples (not all) must be raised to 14 (rather order to get a sharper colour change.

## SODIUM HYDROXIDE NAOH

Dissolve 40 g NaOH in distilled water & make up the volume to 1 lt. Store in polythene bottle.

## PROCEDURE

Take 100 ml. or smaller portion diluted to 100 ml. so that calcium content is about 10 to 20 mg. Add 5 ml. NaOH solution to produce a pH of 12 to 13. Stir & add about 0.2-0.4 g of indicator mixture. Because of the high pH required, titrate immediately after adding alkali & indicator.

Add EDTA titrant slowly with continuous stirring to the proper end point. Carry out a reagent blank using 100ml. distilled water under identical conditions. Standardize EDTA titrant with 25 ml. of calcium standard. Follow the same method as described for the sample.

## CALCULATION

$$\text{Calcium as mg/lg } CaCO_3 = \frac{V_1 - V_2 \times 1000 \times 254}{\text{ml. sample} \times V_3}$$

Where

$V_1$ = is the vol. of EDTA required for sample.

$V_2$ = Vol. of EDTA required for reagent blank and,

$V_3$ = Vol. of EDTA required for 25 ml. of standard calcium.

## EXPRESSION OF RESULTS

Results are expressed as mg/lt. of $CaCO_3$. Round off the result to the nearest whole number.

## PRECISION & ACCURACY

The precision & accuracy of about ±2 to 4% is expected from a good analyst, when the concentration is more than 254 mg/lt. Precision & accuracy also depends on the quantity of the calcium present.

## PRECAUTION

1. Titrate immediately after adding the indicator because it is unstable under alkaline environment.
2. For recognising end point use a colour comparison blank prepared under identical conditions.
3. The final volume for sample & blank should be kept constant.
4. Use deionised distilled water for solution preparation & any dilution necessary.
5. Add same quantity of indicator to sample and blank so that the colour comparison is better.

Chapter -36

# MEASUREMENT OF ELECTRICAL CONDUCTIVITY IN ENVIRONMENTAL SAMPLES

## GENERAL

Conductivity is a measure of the ability of a water sample to carry an electric current and it depends on the ionic strength of the water. This mostly depends upon the nature of the various dissolved ionized substance and their actual and relative concentration and the temperature at which the measurement is made.

Most of the inorganic acids, bases and salts (such as hydrochloric acid, sodium carbonate and sodium chloride) are good conductor. Conversely, most organic compounds such as sucrose and benzene do not dissociate in aqueous solution or pass a current very poorly, if at all.

The determination of electrical conductivity is a rapid and convenient means of estimating the concentration of electrolytes. In waters containing mostly mineral salts (waters used for public supply and many other ground or surface waters), this will not be very different from that or dissolved solids. hence, in such samples, conductivity can be used as a measure of dissolved solids.

For a particular sample one can have a factor to convert electrical conductivity to dissolved solids. This factor will vary from sample to sample. Thus in some cases, measuring the conductivity and converting it into dissolved solids, can be used as a check for the quality of the effluent discharged or for monitoring the treatment plant. The factor usually varies from 0.55 to 0.9. The factor is multiplied by conductivity (μ mhos/c.m.) to obtain dissolved solids. Some other important applications of conductivity measurements are-

1. To purify of distilled and deionised water can be checked by conductivity determination.
2. To check the variations in the dissolved mineral concentration of raw water or wastewater samples quickly.
3. Determination of the amount of ionic reagent required for certain precipitation and neutralization reactions, the end point in such cases is determined by a sudden change in the slope of the conductivity curve.
4. An approximate estimation of the aliquots which may be taken for the chemical determination.
5. Since it is a measure of dissolved solids, it can be used to assess the effect of diverse ions on chemical equilibrium, physiological effects on plant or animal, corrosion rates etc.

The standard unit of electrical conductivity is the ***Siemen per meter*** $S^{-1}M = mhos\ ^{-1}M$) in order to avoid the expression of results in small decimal fractions, a smaller unit the millisiemen per metre ($Msm^{-1}$) is generally used.

Freshly prepared distilled water has a conductivity of 0.1 to 0.2 Ms $M^{-1}$ (or 1 to 4 μ mhos $cM^{-1}$) or even less, but this increases on storing due to absorption of atmospheric carbon dioxide to the extent of 0.2 to 0.4 $msM^{-1}$ or 2 to 4 μmohs $cM^{-1}$ in about a week time. The electrical conductivity of most of the fresh and finished waters is in the range of 5 to 50 ms $m^{-1}$ (or 50 to 500 mhos $cm^{-1}$). The values for highly mineralized water go up to 100 ms $M^{-4}$ and even higher. Some industrial wastes may have conductivity even more than 1000 ms $M^{-1}$.

Electrolytic conductivity increases with temperature at a rate of approximately 2% per degree centigrade. Thus, significant errors can result from inaccurate temperature measurement. It is desirable to make conductivity measurement at 25°C. Additionally, no chemical preservation of sample is permissible.

## APPARATUS

### *Conductometer*

Many instruments of different ranges are available. Most of which are mains operated. Essentially they consist of following two main parts:-

### Conductivity cell

This consists of a pair of rigidly mounted electrodes. Each cell has its own constant depending on its shape and the position, size etc. of the electrodes.

This constant is required to calculate the conductivity of the sample. This can be determined by using a standard solution e.g. 0.01 N KCl. Use Table 36.1 in the conductivity water or by comparison with a cell of accurately known constant.

**Table 36.1 : Conductance of KCl solution at 25°C**

| *Concentration* | *Conductance ms $m^{-1}$* | |
|---|---|---|
| | *Equivalent* | *Specific* |
| 0.0001 | 14,943 | 14.94 |
| 0.001 | 14,695 | 147 |
| 0.01 | 14,127 | 1413 |
| 0.1 | 12,896 | 12980 |
| 1.0 | 11,187 | 111900 |

## INSTRUMENT FOR MEASURING ELECTRICAL CONDUCTIVITY

Most of the instruments consists of a source of alterating current, a wheatstone bridge and a null indicator. Other instruments measure the relation of alternating current through the cell, a voltage across it and have the advantage of a linear reading of conductance. The instrument chosen should be capable of measuring conductivity with an errors not exceeding 1% of 0.1 S $m^{-1}$ (or 1 mhos $cm^{-1}$) whichever is greater.

The conductivity cells should be stored in distilled water when not in use. Cells with platinum electrodes should be replatinized whenever it gives erratic readings.

To platinize, prepare a solution of 1 g chloroplatinic acid (platinum chloride) and 12 mg lead acetate in 100 ml. water. Immerse the electrodes in this solution and connect both to the negative terminal of a 1.5 V dry cell battery. Connect the positive side of the battery to a piece of platinum wire and dip the wire into the solution. The amount of current should be such that only a small quantity of gas is evolved. Continue the electrolysis until both cell electrodes are coated with platinum black.

The platinizing solution may be saved for subsequent use. The electrodes should be rinsed thoroughly and when not in use, keep them immersed in distilled water.

A different type of electrode with a bright surface is also available. Such cells consists of electrodes made from durable metals like stainless steel are widely used for continuous monitoring and field studies.

These cells are calibrated by comparing the conductivity of the water being tested with the results obtained with a laboratory instrument.

**Reagents**

Conductivity meter is used for the preparation of all the standard solutions and this should have a very low conductivity. This must be completely free from carbon dioxide. Redistilled water must be boiled just before use and is allowed to cool. The conductivity of this water should be to less than 0.1 ms $m^{-1}$

## STANDARD POTASSIUM CHLORIDE (0.01 N)

Dissolve 745.6 mg anhydrous KCl in conductivity water and make up to 1000 ml. at 25°C. This is the standard reference solution having specific conductance of 1413 ms $m^{-1}$. It is satisfactory for most waters when using a cell with a constant between 1 and 2. For all other cell constants, stronger or weaker potassium chloride solution listed in Table 36.1 will be needed. Store this solution in glass stoppered pyrex bottles.

## PROCEDURE

1. ***Determination of cell constant***

   Rinse the conductivity cell with at least 3 portions of 0.01 N KCl solution. Adjust the temperature of a fourth portion of $25.0 \pm 1^{o}C$. Measure the resistance of this portion and note the temperature. Compute the cell constant C from equation:-

$$C = \frac{0.001413\ RKCl}{1 + 0.0200\ (t - 25)}$$

2. ***Conductivity measurement***

   Rinse the cell with distilled water followed by portion of sample to be tested. Adjust the temperature of the sample to be tested. Adjust the temperature of the sample of $25 \pm 0.1^{o}C$. Measure the resistance of the sample and note the temperature.

3. ***Calculations***

   The electrical conductivity of the sample can be obtained by multiplying the observed conductivity (in ms $M^{-1}$) by the cell constant. If the instrument measure electrical resistance then divide cell constant by observed electrical resistance :-

$$E = C - EM \quad \text{or}$$

$$E = \frac{C}{EM}$$

To account for the temperature differences, divide the result by the factor (1 + 0.0200 (t - 25) i.e.

$$E = C - EM \quad \text{or}$$

$$E = \frac{C}{EM} \quad \text{or}$$

$$E = \frac{1}{1 + 0.200\ (t - 25)}$$

Where,

E = Electrical conductivity of sample in ms $M^{-1}$

C = Cell constant

EM = Electrical conductivity measured as per meter

t = temperature of sample (°C)

## EXPRESSION OF RESULTS

Express the results in millisiemen per meter (MS $M^{-1}$) Round off the result as indicated in Table 36.2.

**Table 36.2 : Result expression**

| *Range MS $M^{-1}$* | *Record to nearest* |
|---|---|
| 1.0 to 2.0 | 0.02 |
| 2.0 to 5.0 | 0.05 |
| 5.0 to 10.0 | 0.1 |
| 10.0 to 20.0 | 0.2 |
| 20.0 to 50.0 | 0.5 |
| 50 onwards | To the nearest whole number |

## PRECISIONS AND ACCURACY

The precision and accuracy of measurement depends on the instrument used and accuracy with which the cell constant is determined. The precision and accuracy of about ±1-2% is expected with a satisfactory equipment and a good analysis. It is important to note that several conductometers indicate reading as µs $cm^{-1}$ or µmho $cm^{-1}$ This reading is to be divided by 10 to convert it into ms $M^{-1}$. Following conversions may be used :-

10 µs $cm^{-1}$ = 10 µmho $cm^{-1}$ = ms $M^{-1}$

## PRECAUTIONS

1. When electrodes are not being used always keep them in water.
2. Always record the temperature and apply temperature correction. Errors in this may cause significant error in the results.

3. The final conductivity results mainly depend on the cell constant and hence cell constant determination will cause error in conductivity measurement too.
4. Always use conductivity water for preparing standards.
5. Conductivity measurement should be made on the spot or as soon as possible.

## Chapter -37

# MEASUREMENT OF TURBIDITY IN ENVIRONMENTAL SAMPLES

### GENERAL

Turbidity is one of the basic characteristics of water such as colour, odour and potability. It is caused by the presence of suspended and colloidal substances of various origin like clay, silt, mud, silica, rust, calcium carbonate as well as microscopic aquatic organisms like algae, bacteria, planktons etc.

### SIGNIFICANCE

The turbidity being connected with potability of water directly, indirectly or aesthetically, its measurement is very important in deciding the quality of water samples. In fact, the turbidity is taken as an indicator for efficiency and performance of coagulation, flocculation, sedimentation and filtration processes of water treatment plants.

A continuous measurement & recording of turbidity is very helpful in monitoring of streams and water treatment plants. The turbidity is also important consideration in quality control for industries manufacturing analytical grade laboratory reagents and chemicals, cosmetics, food products, beverages etc.

Following values illustrate the above point in Table 37.1 :

**Table 37.1 : Turbidity in aquatic environment**

| *Activity* | *Turbidity in NTU* |
|---|---|
| Raw Water (Rainy Season) | 1000 & above |
| Treated tap water | 1-2 |
| Demineralised water | 0.1-0.5 |
| Water filtered through membrane filter | 0.05 or less |
| Carbon tetrachloride | |
| - commercial grade | 0.05 or more |
| - reagent grade | 0.02 or less |

## MEASUREMENT OF TURBIDITY & TURBIDITY UNITS

1. Based on visual method and,
2. Based on direct meter reading method.

The turbidimeter mainly consists of following 3 parts :

a. Light source (candle or lamp)

b. Sample compartment along with optical parts like lens, mirror and,

c. A device for the measurement of light either transmitted thorough the sample or scattered from the suspended particles in the sample.

Turbidimeters like Jackson candle & Helliage are based on visual methods and use ability of user's eye to detect variation in light intensities, while ach & Fisher turbidimeters are direct reading instruments based on Nephelometric principle.

As the turbidity is an optical instrument depending upon the range of turbidity & size of the particles, it is essential to follow appropriate principle of measurement.

For high (25-5000 units) & medium (5-25 units), turbidities involving large & medium sized particles, it is sufficient to measure transmitted (and or scattered) light. While for measurement of low (less than 1 unit) turbidity, which is generally caused by fine particles (colloidal solutions), transmitted light cannot give required accuracy and sensitivity, & it is essential to use Nephelometric principle of measurement of light scattered of right angle to incident light.

The instrument measuring transmitted light and based upon visual methods, although simple and cheaper, are not suitable for continuous measurement & automatic control.

The accuracy of such instruments is also limited & depends upon the skill of user's eye. The instruments like ach & Fisher based on Nephelometric principle are suitable for lowest turbidities (upto 0.1 unit) and their sensitiviy and accuracy is also quite high because of use of devices like electronic amplifier for measurement of light intensities.

The instrument being direct meter reading are most convenient for continuous measurement recording and automatic control of turbidities.

## UNITS

Turbidity was previously expressed in Jackson Candle Turbidity Units (JTU). This unit is now replaced by more appropriate unit called Nephelometic Turbidity Unit (NTU) because of the use of Nephelometric method of measurement of low turbidities.

## CALIBRATION & STANDARDS

Suspension of Fuller's earth, silica, bentonite, latex kaolin etc. were used for any years for the calibration of a turbidimeter. For mazine is now used as turbidity standard for calibration, because it posses most of the desired properties which other standards do not have.

## JACKSON CANDLE TURBIDIMETER

This was taken as a standard instrument till recent years. Different parts of the instrument are shown in Fig. 37.1 Sample is added in the sample tube till the image of the candle flame when viewed through completely disappear.

The level of the sample in the sample tube which is directly calibrated in turbidity units (JTU) gives the sample turbidity. The operation of this is based on the principle that the light path through a suspension which just cause the image of the flame to disappear. The longer the path lower the turbidity. For the correct operation of the turbidity-meter using above principle, the candle flame must be kept of constant size & kept at constant distance from bottom of the sample tube. The candle must be of bee wax & should burn at constant rate within the limits of 114 to 126 grains per hour.

Candle support should have spring so as to keep the top of the candle pressed against the support. The sample of 21.5 C.M. in depth which just obscures the candle image is said to have turbidity of 100 Jackson Candle Units.

**Interferences**

1. Dirty glasswares.
2. Air bubbles in sample.
3. Vibration that disturb the surface visibility.
4. Colour of sample.
5. Non-uniform rate of burning of the candle exchanges in the intensity.
6. To avoid background light interferences, observation are to be taken in the dark room.

**HELLIAGE TURBIDIMETER**

Sample is taken upto the mark i the turbidity cell & view through the eyepiece. There are two fields. The central field corresponds to the light transmitted through the light scattered from the suspended matter of the sample.

The light intensity of the central spot is balanced with the surrounding field by turning the dial on the right hand side of the instrument. This dial contfols the brightness of the central field.

In general, there is definite interval over which the two light fields appear as balanced. Hence, the dial is rotated from balanced. Hence, the dial is rotated from lower values towards high values until the black spot in the centre of the field just disappear. This instrument is also based on visual method & hence possesses all the disadvantages mentioned earlier.

## HACH LABORATORY TURBIDIMETER

This is the instrument based on purely Nephelometric method for the measurement of turbidity of liquids & operated on the principle that the light passing through a substance is scattered by particulate matter, suspended in the solution. The intensity of the scattered light serves as a measure of the turbidity.

As shown in Fig. 37.2, a beam of strong light is sent upward through a cell containing the sample. As the beam passes through the suspended particles, an amount of light (directly proportional to the turbidity) is scattered at right angles (90°) to the beam & is received by a photomultiplier tube. The light energy thus converted into an electrical signal is further amplified using an amplifier & then measured by a meter.

## MAINTENANCE

Cleanliness is of utmost important in all turbidity measurements. All the glass and optical parts must be protected from dust & scratches. Even the finger marks on the sample tube bottom, mirror, source lamps & reflectors may disturb measurement, accuracy & calibration.

To prevent the polished bottoms of the sample tubes & plungers from scratching it is advisable to keep these items on a soft cloth when not in use.

The sample tubes & condensing lenses etc. may be washed with mild detergent & rinse with distilled water.

The filters & mirrors may be cleaned using a soft tissue paper or air syringe & may be wiped occasionally with a soft tissue soaked with distilled water or mild detergent.

Never immerse these items in water or cleaning solution.

Always wipe the outside of the sample tube before placing into the holder otherwise it may keep the mirror wet for long time & damage it permanently.

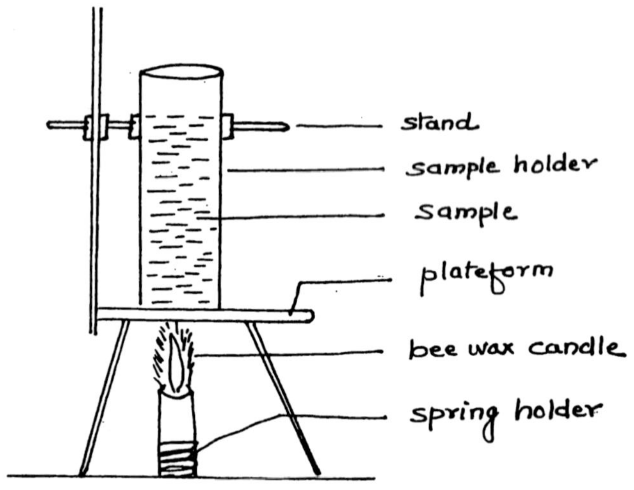


**Fig. 37.1 : Jackson candle turbidimeter**

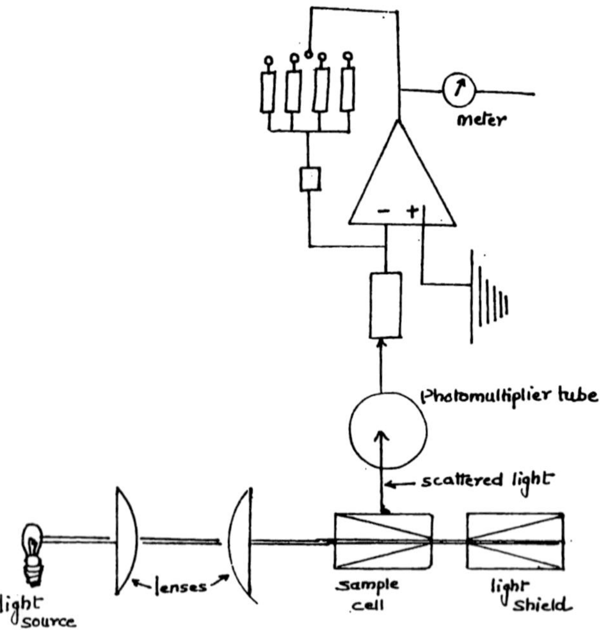


**Fig. 37.2 : Schematic circuit diagram of Hack turbidimeter**

# Chapter -38

# INFRA-RED SPECTROPHOTOMETRY

## INTRODUCTION

In 1905, Cobentz has done the IR studies on a large number of compounds & further found out that the selective absorption of IR radiation arise out of vibration of atoms in a molecule. When the Infrared radiation falls on the substance, the certain amount of energy is absorbed by the molecules & the transition in vibrational & rotational energies take place. Thus I-R spectrum of a substance is a plot of absorption versus the frequency falling on the substance & is very much characteristic of a substance.

Let us take a very simple molecule having only two atoms A & B. If now the infrared radiation a falls on this molecule, the vibration which is a periodic stretching along the axis, A-B occurs just as in case of a spring being stretched & then left. I-R is thus more favourable for organics since there is light dependence on the aggregation of molecules & so leads to vibration frequencies but this is not so far the inorganic because of the two factors-

1. Water is the most common solvent that absorbs most of I-R.
2. The bands are not sharp & very wide.

Thus, I-R spectroscopy now-a-days is employed to a great extent in industries, in organic chemistry for identification of functional groups which are taken as the structural units in as unknown compound.

## PRINCIPLE

The presence of an absorption band is no doubt an indication of possible structural, but absence of any band in a chosen region is of immense help in stating those functional groups & can be eliminated.

The various types of linkages that can be studied by I-R can be classified as under—

1. Carbon-Carbon (C-C) & Carbon-Hydrogen (C-H)
2. Carbon-Oxygen (C-O) & Oxygen-Hydrogen (O-H)
3. Carbon-Nitrogen (C-N) & Nitrogen-Hydrogen (N-H)
4. The last group includes the linkages which relate to the inorganic structures: i.e. is sulphur-

   C-F

   C-S.

   C-C[1].

   P-O.

   Si-C.

## INSTRUMENTATION

In every I-R instrument, there is a source that provides the radiation over the complete I-R range, then, there is a monochromator that disperse this energy & selects out a narrow frequency range. Finally the energy is measured by the detector which transforms the I-R heat energy into electrical signal. Here the prism has to be turned manually & the measurements were done at different frequencies in I-R range. The output was measured with the help of precalibrated potentiometer in terms of optical density on obtaining the null.

This way many complications of the instrument were avoided, especially the linearity of the galvanometer & then that of amplifier is only necessary. This method is very useful especially when the quantitative measurements are done at a particular I-R frequency. Here, with respect to different concentrations of substances what is done is to adjust the null & read the potentiometer reading.

Now-a-days, the entire spectrum is automatically recorded & measurements can be done from this, a chart can be obtain. Such an instrument depicted diagrammatically in fig 38.1.

Here the source is a heated high resistance silicon carbide rod which attains the temperature between 1600-1800°C. This provides the radiation over the complete I-R range. This is dispersed & selected by the monochromator & falls on the detector. $M_1$ & $M_2$ are metallic mirrors for reflecting energies into two beams, Sample beam reflected by mirror $M_3$ & then falls on the rotating mirror & again reflected, while beam passes straight. Ultimately both the focused on the detector D. When the energy of the beam passing through sample & reference is equal, there is no signal. If any absorption in sample beam occurs, then only signals are produced & recorded.

## DETECTOR

Unlike UV & VIS, here the energy in I-R is heat rather than light. So the detectors are different from photo tube & photomultipliers. In many instruments, thermocouple and an amplifier is used.

## SAMPLE TECHNIQUE

Only single state I-R spectrum of the substance does not give the unmistakable information. For this, we must get the spectra of the substance by either dilution or changing concentration or by changing the phase of substance i.e. by taking liquid phase, solid phase, gaseous phase.

The spectra of the substance in more than one phase will only confirm about the functional group position. Sample preparation is a difficult task & many times, we do not get the good results. In the study of solid samples by null the film has to be as thin as possible to avoid general absorption.

If the sample is crystalline, there is more energy loss due to scattering. The scattering can be reduced to a great extent by taking the sample in per fuiro-kerosene or paraffin oil or hexachlorobutadine. Then mixing this thoroughly in agate pestle & mortar & spreading it on the NaCl plate to a thin film.

But such a spectra has a drawback that the specta can not be reproducible exactly & is not useful for quantitative purpose. In the solid state spectra itself there is another sampling teqhnique known as the disc technique. Here the substances finally powdered & meshed to 300 mech, about 6 mg is mixed with 200 mg of KBR of the same above mesh.

The mixing is done by the vibration mill & then this is put in stainless stell disc & pressed to 20 tons/inch.[2] Thus a transparent disc with 10 mm diameter & about 1 mm thickness is formed.

This then is inserted in the sample beam & the spectra can be recorded. This also has some disadvantages.

## SOLUTION SPECTRA

The solution spectra is more reliable but at the same time, more complicated because of the limited choice of the organic solvent which do not disturb throughout the wide range of the IR. Generally $CCl_4$, $CS_2$ are used in combination.

## SELECTION OF SUITABLE SOLVENT

The best solvent for the I-R work can be selected on the following characteristics :

1. It should have a spectrum relatively free from absorption bands.
2. It should have good solubility.
3. It should, though, dissolve the substance, should not reach with it, in absorption cells.

The last one is the most important as the cells in I-R spectrophometry need the special mention & care in handling as there are made out of the transparent salt plates such as NaCl, KBr, $CaBr_2$ etc. Thus in the choice of solvents when we are studying the OH group, $CCl_4$, $CS_2$ or even $CHCl_3$ will be most convenient solvents.

$CCl_4$ and $CS_2$ as organic solvents can always be used in combination. They have as stated above a high transmission in infrared range & where one absorbs the other transmits.

## PEAK SELECTION

This is done by spectral differences between the various compounds & it is observed that the region 7.5 M to 14 M is the most suitable for this purpose. For here each substance has different characteristic spectrum. This is why many times, this is called the finger print region of the substances. This is the region of infra red where the substances can be idenfied, studied & finally measured. Infrared of organics is a characteristic of the molecule & it must give rise to frequencies in combination of atoms.

## APPLICATION

It can be seen that the I-R spectrum is a plot of percent transmission versus the frequency positions furnish the data for the qualitative analysis while percent transmission values furnish the data for the quantitative analysis.

Thus, we see that the I-R spectrum of a substances will be most useful as under:-

1. ***Identification of Organic compounds :***

   This is most easily done by comparison with the standard pure substances or on following the standard punch card spectra.

2. Detection & identification of small amount of impurities in organic compound. This is the most difficult but an important task performed by I-R techniques.

   Here the spectra of the compound can be suitably taken in more than one phases & the spectra of all the substances in all phases (liquid, solid, gases) do not differ. Many a times the presence of impurities is deleted in a best way by putting the pure substances in the references cell & the one under test as the sample one. This is known as ***differential analysis*** & serves in a best way to find the impurities. If the substance is to be purified then after every purification process the above test is done till the band due to impurity vanishes.

3. The accurate quantitative determination of these impurities

## ADVANTAGES

The following are some main advantages for the IR technique:

1. The amount of sample required by this is a very small, often it is injected by syringe.

2. No preliminary refinement required.
3. Method is most sensitive to small concentration.
4. It is accurately quantitative
5. This requires very short time complete the analysis.

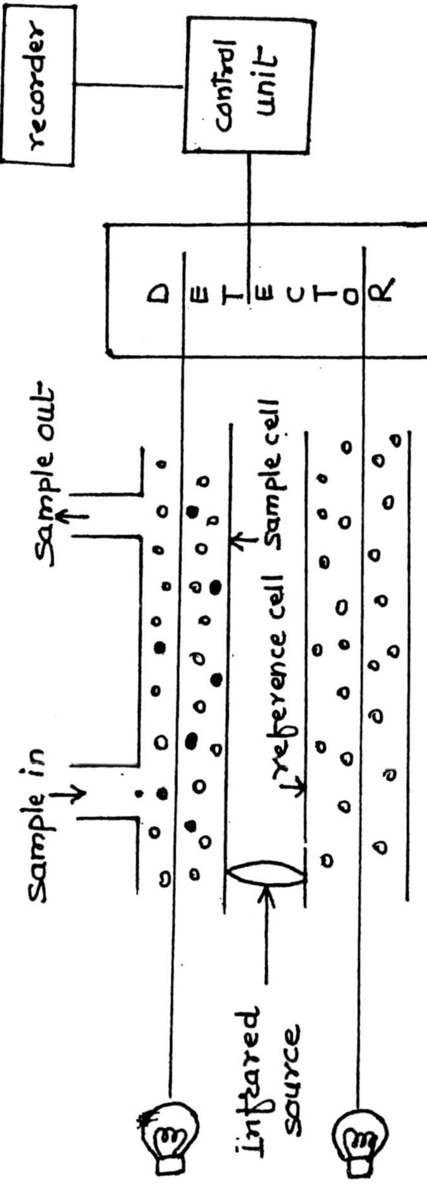


Fig. 38.1 : Schematic diagram of a typical non-dispersive infrared spectrophotometry

# Chapter -39

# MEASUREMNT OF DISSOLVED OXYGEN IN THE ENVIRONMENTAL SAMPLES

## GENERAL

Dissolved oxygen (D.O.) in natural & waste water depends on the physical, chemical & biological activities in the water bodies. The analysis for D.O. is important in water pollution control & wastewater treatment. D.O. as a parameter is used to :-

1. Assess quality of surface water & to keep a check on strcam pollution.
2. It is the basis for BOD which is an important parameter to evaluate pollution potential of wastes.
3. It is responsible for corrosivity, photosynthetic activity & septicity.
4. D.O. is necessary for all aerobic biological wastewater treatment process.

## PRINCIPLE

**The iodometiric test is precise & reliable titrimetric procedure for D.O. analysis. The test is based on the addition of divalent maganese solution, followed by strong alkali, to the sample in a BOD glass-stoppered bottle D.O. present, rapidly oxidise an equivalent amount**

of the dispersed divalent managanous hydroxide of higher oxidation state (brownish-orange in colour).

This brownish-orange colour indicates presence of oxygen. In the presence of iodide & on subsequent acidification, higher manganese hydroxide revert to the divalent state & liberate iodine equivalent to the original dissolved oxygen content of the sample. The iodine liberated is titrated with a standard solution of sodium thiosulfate using starch as an indicator.

## INTERFERENCE

It is caused by certain oxidizing agents, which liberate iodine from iodides such as ozone, chlorine, ferric compounds, manganese in tri or higher oxidation states chromate, nitrate, persulphate, peroxides etc., by certain reducing agents which may reduce iodine to iodide (such as ferrous compounds, thiosulphate, readily oxidizable organic matter, nitrite etc.

Some organic compounds hinder the settling of the precipitate & colour of some may interfering the starch end point. Modification are included for minimize the effect of interfering materials.

## APPARATUS

The bottle used should be of good quality having 300 ml. capacity with narrow neck & well fitting ground glass stopper. It is convenient to have number on bottle & their stopper.

## REAGENTS

a) ***Manganoius sulphate solution***

Dissolve 500g $MnSO_4$ or 480 g $MnSO_4$ $4H_2O$ or 400 g $MnSO_4.2H_2O$ or 364 g $MnSO_4$. $H_2O$ in distilled water. Filter & dilute to 1 lt.

This manganous sulphate solution should not give a colour with starch when added to an acidified solution of potassium iodide.

b) ***Alkali iodide - azide reagent***

Dissolve 500 g NaOH (or 700g KOH) and 135 g NaI (or 150 g KI) in distilled water & dilute to 1 lt. Add 10 g sodium azide

($NaN_3$) and dissolve in 40 ml. distilled water. Potassium & sodium salts may be used interchangeably.

This reagent should not give colour with starch solution when diluted & acidified.

c) ***Sulphburic acid***

Conc. $H_2SO_4$ - one ml. of this is equivalent to about 3 ml. of alkali - iodide - azide reagent

d) ***Starch indicator***

Make a smooth paste of 5 g soluble starch in cold water & pour this into 1 lt. boiling water with constant stirring. Boil for one minute & allow to cool before use.

Use clear supernatant Pressure. with 1.25 g saliclylic acid per litre or by adding a few drops of toluene.

e) ***Sodium thiosulphate stock dissolution* (0.125 N)**

Dissolve 31.5 g sodium thiosulphate pentahydrate ($Na_2S_2O_3$ $5H_2O$) in 1 litre copper free boiled & cooled distilled water. Add 5 ml. chloroform or 1 g NaOH as preservative.

f) ***Standard Sodium thiosulphate titrant* (0.0125 N)**

Dilute the stock sodium thioslphate solution to obtain 0.0125 (100ml of stock 0.125 N ($Na_2S_2O_3$) diluted to 1 lt. with distilled water give 0.0125 N ($Na_2S_2O_3$). Standardize this diluted solution daily.

***Standardization***

Pipette 10 ml. 0.0125 N Potassium iodate into a conical flask containing about 100 ml. distilled water Add 2 ml. of conc. $H_2SO_4$ followed by about 2 g of potassium iodide solid.

Titrate immediately against sodium thiosulphate, 10/25 N, using starch as indicator. Apply concentration correction in the calculation.

***Potassium Iodate (0.0125 N)***

Dissolve 0.446 potassium iodate previously dried at about 120°C in distilled water & dilute to 1 lt. the solution is stand for a long period if stored in a glass-stoppered bottle.

## PROCEDURE

1. Add 2 ml. manganous sulphate followed by 2 ml. alkali-iodide-azide reagent, dripping the pipettee little below the surphase.
2. Stopper carefully to exclude air bubbles & thoroughly mix the contents. To attain this invert the bottle at least 15 times.
3. All the precipitate to settle to the lower one third of the bottle. Repeat the mixing & allow the precipitate to settle completely leaving a clear supernatant liquid.
4. Add 2 ml. conc. $H_2SO_4$ immediately after removing the stopper. Restopper & mix by gentle inversion until all the precipitate dissolves. If it is does not, allow to stand fora few minutes & repeat the mixing,
5. Distribute the iodine uniformly before taking a part of it titration.
6. Take 100 ml. of sample for titration (1.3 ml. extra to account for 4 ml. or the regents, manganous sulphate & alkali-iodide-azide, added to 300 ml. of sample) in a conical flask & immediately titrate the liberated iodine with the standard thiosulphate solution to a pale yellow straw colour.

   Add 2 ml. starch solution & continue the titration till the first disappearance of the blue colour. Disregard subsequent recoloration due to the catalytic effect of nitrite of ferric salts.

   It must be remembered that iodine is volatile & therefore the titration must be carried out as expeditiously as possible & with the minimum of exposure to the air.

## RIDEAL-STWEART MODIFICATION FOR ELIMINATION OF INTERFERENCE BY FERROUS IRONS

This modification is used when the sample contains ferrous iron. Add 0.7 ml. conc. $H_2SO_4$ followed by 1 ml, potassium permagnate (0.63%) & 1 ml. potassium fluoride (40%) to the sample bottle, stopper & mix by inversion. The amount of permanganate added should be just sufficient to obtain a violet tinge that persists for 5 minutes.

Remove the excess per manganate by add 0.5 to 1.0 ml. potassium oxalate (2%). Mix well & let stand in the dark to facilitate the reaction. Excess oxalate causes low results & hence add only an amount of oxalate that completely decolorizes the potassium permanganate.

After applying this modification, usual method should be followed:-

## CALCULATION

$$\text{Dissolve Oxygen content mg/l} = \frac{\text{Vol. of thiosulfate required for 101.3 ml sample x 10}}{\text{Vol. of thiosulfate required for 10 ml. of 0.0125 N potassium iodate.}}$$

**Vol. of thiosulphate required for 10 ml. of 0.0125 N** potassium iodate.

## EXPRESSION OF RESULTS

Express the dissolved oxygen **results in mg/lt. Report** the values to the nearest of first decimal i.e. 0.1 mg/lt.

## PRECISION & ACCURACY

1. Do not use scap or synthetic detergent for washing **the D.O.** bottles.

   Normally bottles are kept clean by the acidic iodine **solution of** winkler procedure & require no further treatment except **through** rinsing with the tap water.

2. New bottles should be cleaner with 5 N sulphuric or hydrochloric acid & then rinsed thoroughly with the tap water.

3. Just before taking he sample rinse sample bottle with the sample.

4. Bottle should be completely filled with the sample & remove bubbles entrapped into the bottle while taking sample by tapping the neck of the bottle with the stopper.

5. Fix the D.O. on the site by adding manganous sulphate & alkaliodide-azide solution.

6. The titration should be carried out as expeditiously as possible & with minimum exposure to the air, so as to avoid any losses of iodine during titration.

7. Use BOD - sample discharger to avoid inconvenience caused in measuring 101.3 ml.

This will save time as well as maintain uniformity in the comparative analysis.

8. Avoid errors due to carelessness in collecting the samples, prolonging the completion of the test or due to any interferences.

## Chapter -40

# MEASUREMENT OF VARIOUS TYPES OF RESIDUES OR SOLIDS IN ENVIRONMENTAL SAMPLES

## GENERAL

A highly mineralized water is less acceptable than a water of little or moderate mineral content for drinking, household and special industrial purposes. A good indication of measuring the quantity of minerals present in water or wastewater can be determined by finding out the mineral left after evaporation.

The term residue applied to the material remaining in the dish after evaporation and drying of a water sample in an oven at a definite temperature. The terms ***'Suspended Solids'*** and ***'Dissolved Solids*** correspond to ***nonfilterable*** and ***filterable residue***, respectively. ***'Fixed Solids'*** and ***'Volatile Solids'*** correspond to after ignition and loss due to ignition at a definite temperature respectively. (Table 40.1)

## TOTAL RESIDUE (TS)

### Principle

A well mixed sample is evaporated and dried in a weighted china dish in an oven at 105 ±1°C. The increase in weight represents the total residue.

**Table 40.1 : Types of residue**

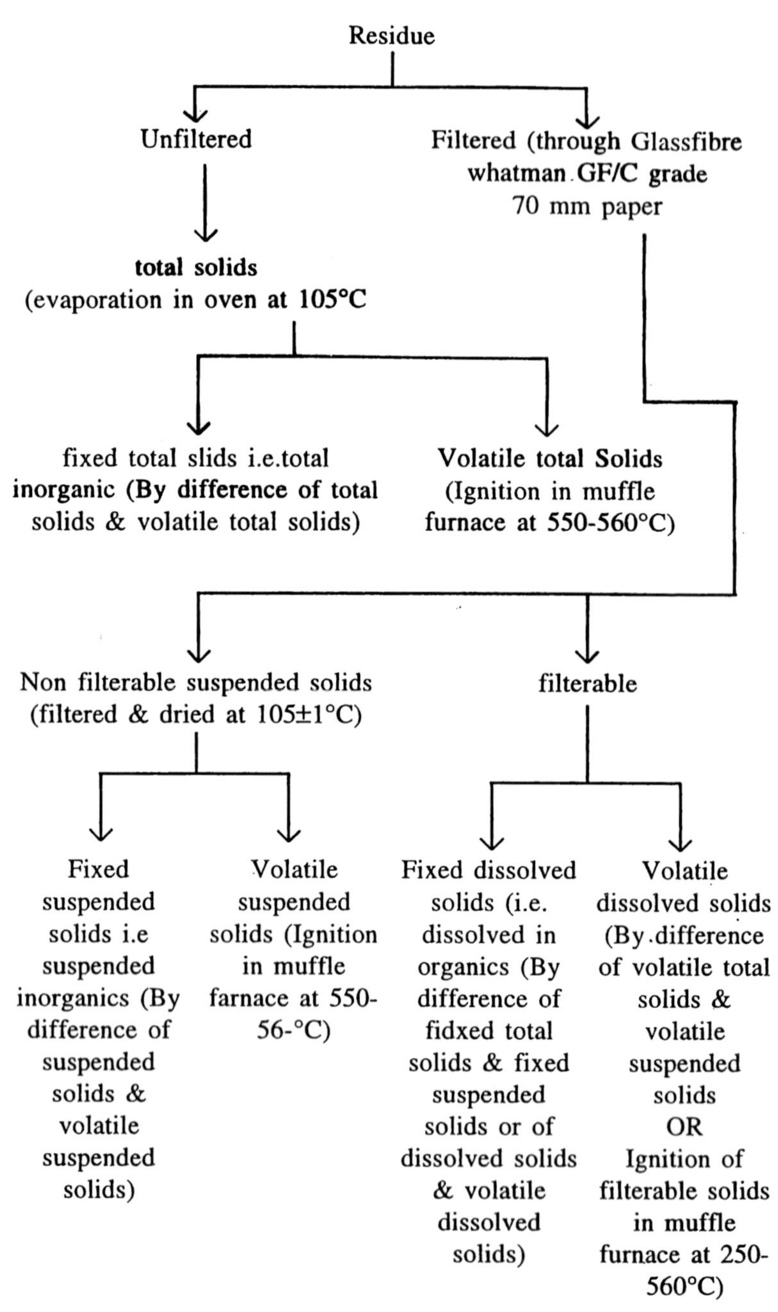

## APPARATUS

1. Evaporating dishes of 150-200 ml. capacity made of porcelain. Alternatively beakers can be used.
2. Drying oven at 105° ±1°C
3. Desiccator provided with a desiccant.
4. Analytical balance capable of weighin upto 0.1 mg.
5. Muffle furnace at 550-560°C

## PROCEDURE

1. Heat the clean evaporating dish at 550 ± 50°C for 1 hr. in a **muffle furnace.**
2. Cool, desiccate, weight and store the dishes in a desiccator until it is used. On repetition of this process dish should show a constant weight.
3. Transfer a known volume of well mixed sample to the pre-weighted dish and evaporate to dryness in the oven at 105±1°C. Choose a sample volume in such a way that it will yield a reduce between 25 to 250 mg and preferably between 100 and 250 mg (usually 50 to 100 ml. for wastewater and 100 to 200 ml. for clean waters is suitable.
4. After complete evaporation dry it atleast for 1 hrs. in the oven.
5. Cool the dish in a desiccators and weigh it.
6. Repeat the cycle of drying, cooling and weighting till the weight of the dish becomes constant.

## CALCULATION

$$\text{Total residue (mg/lt)} = \frac{(x\text{-}y) \times 10^6}{V}$$

Where,

x = wt. of dried sample + dishing

y = wt. of empty dish in g

v = volume of sample in ml.

## EXPRESSION OF RESULTS

Report the result as total residue or total solids in mg/lt. Round off the results to the nearest whole number.

## PRECISION AND ACCURACY

The precision is about ± 4 mg or ± 5% for raw waters, the results are considered statistically unreliable due to sampling errors.

## FIXED SOLIDS AND VOLATILE SOLIDS

### Principles

The residue produced on evaporation (at 105±1°C in oven) is ignited in a muffle furnace at 550-560°C. The remainder and the loss on ignition represent fixed and volatile solids respectively.

## PROCEDURE

1. Same apparatus is used for fixed - and volatile solids measurements as used in Total Solids estimation. After completion of the test for total residue, ignite the dish with the residue in a muffle furnace at 550-560°C for about 30 minutes
2. Allow the dish to cool partially in air. Most of the heat dissipated or switch off the muffle furnace and let the temperature fall.
3. Transfer the dish to a desiccator, cool and weight it as soon s it cools completely.
4. Repeat the operation of ignition, cooling and weighing until a constant weight is obtained (i.e. till the difference between two successive series of operation is not more than 0.5 mg or 1% of the remainder solids).

## CALCULATION

$$\text{Volatile residue, mg/lt.} = \frac{(x - y) \times 10^6}{V}$$

Where,

x = weight of sample + dish (before ignition but after evaporation in gms).

y = Weight of sample + dish after ignition in gms.

z = weight of dish in g

v = Volume of sample in ml.

## EXPRESSION OF THE RESULTS

Repeat the results as 'fixed residue' or fixed solids and volatile residue or volatile solids in mg/lt. Round off the results to the nearest whole number. Precision and Accuracy are same as mentioned in Total Solid estimation.

## DISSOLVED SOLIDS

Dissolved solids or filterable residue, are those solids capable of passing through a standard glass fibre filter paper. In water containing no suspended solids, the dissolved solids are equal to the total solids). The filtrate from the nonfilterable residue or suspended solids can be used to determine the filterable residue.

Fixed dissolved solids and volatile dissolved solids correspond to the filterable solids remainder after ignition and loss by ignition, respectively

## APPARATUS

1. Glass fibre filter paper, Whatman GF/C grade or equivalent.
2. Millipore filtration assembly with suction flask of 500 or 1000 ml. capacity.
3. Porcelain dish or beaker, 150-200 capacity ml. respectively.
4. Drying oven, $180^{\circ} \pm 2^{\circ}C$ $105 \pm 1^{\circ}C$.
5. Desiccator.
6. Analytical balance capable of weighing upto 0.1 mg.

## PROCEDURE

1. Preparation of evaporating dish, heat the clean dish to 550-560°C for about 1 hrs. in a muffle furnace. Cool, partially in air then in desiccator. Weigh immediately before use.
2. ***Preparation of glass fibre disc*** : Place the disc on the membrane filter holder. Wash it with three successive 20 ml. volumes of distilled water. Discard all these washings.

3. Under vauum filter, 100 ml. or more of well mixed same through glass fibre filter.

4. Transfer 100 ml. (or a larger volume if total filterable residue is low) filtrate to a preweighted evaporating dish.

## DISSOLVED SOLIDS AT 105°C

Evaporate and dry the sample in a dry oven at 1052±1°C (dry atlest for 1 hrs after complete evaporation). Cool in a desiccator and Weigh. Repeat the drying cycle unit a constant weight is obtained (i.e. ± 0.5 mg or ± 1% of the evaporated sample)

## DISSOLVED SOLIDS AT 180°C

The above procedure to be carried out at 180± 2°C instead of 105±1°C.

## CALCULATION

$$\text{Total dissolved solids, mg/lt} = \frac{(x\text{-}y) \times 10^6}{V}$$

Where

x = wt. of dried filtered sample + dish in gms.

y = weight of empty dish ingm

v =. volume of sample in ml.

## EXPRESSION OF RESULTS

Express the results as total dissolved Solids dried at particular °C, mg/lt.

## VOLATILE AND FIXED DISSOLVED SOLIDS

## PROCEDURE

1. Determine the proportion of matter volatile on ignition by igniting the dried residue (after during dissolved solids) in a muffle furnace at 550°C for about an hour.

2. Cool in desiccator and weight it

## CALCULATIONS

$$\text{Volatile dissolved solids, mg/lt} = \frac{(x\text{-}y) \times 10^6}{V}$$

Where

x = wt. of dried residue + dish (g) (before ignition)

y = wt. of ignited residue + dish (g) after ignition)

v = volume of filtrate taken in ml.

$$\text{Fixed dissolved solids, mg/lt} = \frac{(y\text{-}z) \times 10^6}{V}$$

Where

y = Weight of ignited residue + dish, g

z = weight of dish g

V = volume of filtrate taken, ml.

## EXPRESSION OF RESULTS

Express results as 'Volatile dissolved solids' and 'fixed dissolved solids' as mg/lt. Round off the results to the nearest whole number.

## SUSPENDED SOLIDS

### Total suspended solids

Suspended solids are those solids which are retained on the standard glass filter disc.

### Sampling handing and preservation

Non homogeneous particulate matter such as leaves, sticks, fish and lumps of fecal matter should be excluded from the sample.

Preservation of the sample is not practical, analysis should be carried out as soon as possible.

## APPARATUS

1. Glass fibre filter disc, whatman GF/C grade or equivalent, 70 mm.
2. Millipore filtration assembly with suction flask.
3. Drying over having temp. 105±1°C
4. Desiccator
5. Analytical balance capable of weighting to 0.1 mg.

## PROCEDURE

***Preparation of glass fibre filter disc***

1. After numbering the filter disc, dry it in an over at 105 ± 1°C for about an hour. cool in a desiccator and weigh it.
2. After weighing place it on the filter holder. Apply vacuum and wash it with about 20 ml. of distilled water.

While the pump is on transfer 100 ml. or an appropriate volume of the well mixed sample so that it contains about 25 to 200 mg of suspended solids, to the filtration of the millipore assembly. If the solids are very high, add small volume in number of instalments.

After complete filtration of the sample add about 20 ml. distilled water.

Transfer the filter paper carefully to the petridish when it is completely dry. Dry at least one hour at 105 ± 1°C in the oven. Cool in a desiccator and weigh.

Repeat the drying cycle until a constant weight is obtained.

## CALCULATION

$$\text{Total suspended solids mg/lt} = \frac{(x\text{-}y) \times 10^{66}}{V}$$

Where

x = wt. of filter paper + residue in g

y = wt. of filter paper g

V = volume of sample, ml.

## EXPRESSION OF RESULTS

Express the result as total suspended solids in mg/lt. Round off to the nearest whole number.

## VOLATILE AND FIXED SUSPENDED SOLIDS

## PROCEDURE

1. After determining the total suspended solids the filter disc containing the suspended matter is ignited in the muffle furnace at 550°C for about on hour.
2. Cool in a desiccator and weight

## CALCULATION

***volatile suspended solids*** mg/lt. $= \dfrac{(x - y) \times 10^6}{V}$

Where

x = weight of dried residue + filter disc ing (before ignition)

y = weight of ignited residue + filter disc ing (after ignition)

v = volume of sample filtered in ml.

***Fixed suspended solids,*** mg/lt. $= \dfrac{(y-z) \times 10^6}{V}$

Where

y = wt. of ignited residue + filter disc ing

z = weight of filter disc in g

v = volume of sample filtered in ml

## EXPRESSION OF RESULTS

Express the result as volatile suspended solids and fixed suspended solids in mg/lt. Round off the value to the nearest whole number.

## GENERAL PRECAUTIONS

1. Always set the balance to zero before use.
2. Before eighing ensure that dish or filter disc is cooled to room temperature.
3. At all stages prevent disc from dust etc.
4. Greater the number of replicates more in the accuracy. Keep atleast 3 replicates.
5. Mark the number on the dish or filter disc before starting the test (before drying).

## Chapter -42

# MEASUREMENT OF BIOCHEMICAL OXYGEN DEMAND (BOD) IN ENVIRONMENTAL SAMPLES

### GENERAL

Biochemical oxygen demand (BOD) is defined as the amount of oxygen required by microorganisms to stabilize biological decomposable organic matter in a waste under aerobic conditions. The biochemical oxygen demand is an approximate measure for the amount of biochemically degradable organic matter present in the sample.

The BOD test is widely used to determine :

1. The degree of pollution in lakes & streams at any time & their self purification capacity.
2. the pollution load of wastewaters &,
3. efficiency of waste treatment plants.

Since the test is based on mainly bioassay procedure measuring amount of oxygen consumed by bacteria while stabilizing organic matter under aerobic conditions, it is necessary to provide standard conditions of nutrient supply, pH, absence of substances inhibiting microbial growth & temperature . Because of the low solubility of oxygen in water, the sample is diluted proportionately depending upon the expected BOD so that the demand does not exceed the available oxygen. A mixed group of organisms should be present in the sample, if not the sample has to be seeded artificially. Temperature is controlled at 20°C. the test is carried out for 5 days as 70-80% of the BOD is satisfied during this period.

## INTERFERENCES & INADEQUACIES

The pH of the sample should be adjusted by adding alkali or acid, in the range of 6.5 to 8.5.

Some samples may be sterile & will need seeding. The purpose of seeding is to introduce into the sample a biological population capable of oxidizing the organic matter. Where such microorganisms are ready present as in surface water, domestic sewage or unchlorinated effluents etc. Seeding is not required.

For seeding, wherever necessary use settled domestic sewage which has been stored at 20°C for 24 hrs., or use fresh sewage. For seeding use 1 to 2 ml. or settled seed to each litre of dilution water. Other interference in BOD test are the same as in the COD test. In addition, lack of nutrients in dilution water, lack of an acclimated seed organisms & presence of heavy metals or other toxic materials such as residual chlorine are other source of interference in this test.

## APPARATUS

## INCUBATION BOTTLES

300 ml. capacity with ground glass stoppers. New bottles should be cleaned with either 5 N hydrochloric or sulphuric acid followed by rinsing with distilled water.

In normal use bottles once used for winkler procedure does not required any treatment apart from thorough rinsing with tap water & distilled water.

As precaution against drawing air in the dilution bottle during incubation, use special BOD bottles & add water to the flared mount of the bottle.

## REAGENTS

### Dilution water

Distilled water alone is unsatisfactory as a diluent & it is recommended that a synthetic dilution water be employed. This can be prepared by adding reagents to good quality of distilled water. Water from copper stills should not be used since copper inhibits biochemical iodation.

Glass distilled or deionized water produced in some commercial units may be used. Stocks solution of the following pure chemicals are required. Any solutions of precipitates or growth should be discarded.

### Phosphate buffer stock solution

Dissolve 42.5 g potassium dihydrogen phosphate ($KH_2PO_4$) in 700 ml. water & add 8.8 gms. sodium hydroxide. This should give a solution of pH 7.2 which should be checked. Add 2. gms. ammonium sulphate & dilute to 1 lt.

### Magnesium sulphate solution

Dissolve 25 g magnesium sulphate $MgSO_4$ $7H_2O$ in some amount of distilled water. After complete dissolve, add and make up volume upto 1 lt. by distilled water.

### Calcium chloride solution

Dissolve 27.5 g anhydrous calcium chloride $CaCl_2$ in 1 lt. distilled water.

### Ferric chloride solution

Dissolve Y. 125 g ferric chloride ($FeCl_3$. $6H_2O$) in 1 lt. distilled water.

### Seeding

The purpose of seeding is to introduce into microbiological population capable of oxidizing the organic matter in wastewater. Samples in which such microorganisms are already present, as in domestic sewage & surface water, seeding is not required.

When the sample contains very few microorganisms, the standard seed material used is sewage either that has been stored at 20°C for 24-36 hrs. The seed added should produce a seed correction of atleast 0.6 mg/lt. Usually 2 ml. to 5 ml. of sewage is used per litre of dilution water.

## PROCEDURE

### Preparation of dilution water

Aerate the required volume of distilled water in a container by bubbling compressed air for 1-2 days to attain D.O. saturation. After aeration, keep it at 20°C for atleast one day.

At the time of use, 1 ml. each of the phosphate buffer, magnesium sulphate, calcium chloride & ferric chloride for each litre of dilution water. The trend of adding these nutrients should be followed as given above. If required, add requisite quantity of seed also. Mix well.

**Dilution of sample**

Neutralize the sample to pH around 7.0 using alkali or acid (NaOH or $H_2SO_4$). The sample should be free from residual chlorine. If it contains residual chlorine too, remove it by using $Na_2SO_3$ solution as given below:-

Take 50 ml. of the sample & acidify with 10 ml. of 1 + 1 acetic acid. Add about 1 gm of KI. Titrate with sodium sulphite using starch as indicator.

Calculate the volume of $Na_2SO_3$ required per ml. of the sample & add accordingly to the sample to be tested for BOD.

Atleast two dilutions of the sample are made so that the depletions are in the range of 40-70%. It is recommended to have lower dilution to give 40% & higher one to give 70% oxygen depletion.

The dilutions may be made as follows:

Requisite quantity of sample is taken in one litre capacity volumetric flask. Dilute this by shipping out the dilution water to the mark. Mix well. Rinse BOD bottles with the sample & then fill the bottles with the diluted sample. As far as possible avoid entrapping of air bubbles in the bottle. Tap the neck of the bottles to ensure the removal of the air bubbles in the bottle. Tap the neck of the bottles to ensure the removal of the air bubbles from the bottle. Stoper the bottle immediately. Keep one bottle for determination of initial (zero day) dissolved oxygen & incubate two bottles at 20°C for 5 days. The bottles incubated should be added daily as some water is lost due to evaporation.

Prepare four blanks by siphoning out only dilution water into the bottle. These bottles should be first rinsed & then filled with the dilution water. Two of these blanks should be used to determine initial dissolved oxygen & rest of the two are incubated for 5 days at 20°C. Usually depletion due to blank in 5 days should not more than 0.2 mg/lt. Dissolved oxygen, initial & after five days is determined as D.O. measurement.

**Seed correction**

If the dilution water is seeded, do not use the seeded blank for seed correction because the 5 day seeded dilution water blank is subject to erratic oxidation due to very high dilution of seed, which is not the oxygen depletion of the seed by setting up a separate series of dilution & selecting those resulting in 40 to 70% oxygen depletion in 5 days. Use one of these or average of these depletions to calculate the correction due to small amount of seed in the dilution water.

## CALCULATIONS

1. When BOD has been determined in an undiluted sample.

   BOD mg/lt =D.O. before incubation in mg/lt. D.O. after incubation in mg/lt.

2. When BOD has been determined in a diluted sample, then -

   $D_1$ = D.O. of diluted sample on 0th day

   $D_2$ = D.O. of diluted sample on 5th day

   $B_1$ = D.O. of dilution blank on 0th day

   $B_2$ = D.O. of dilution blank on 5th day

   $C_1$ = D.O. of seed control on 0th day

   $C_2$ = D.O. of seed control on 5th day

   F = ratio of seed in sample to seed in control

   $$F = \frac{\%\ \text{seed in } D_1}{\%\ \text{seed in } C_1}$$

   Seed correction = $(C_1 - C_2)\,F$

   BOD without applying seed correction :

   $$\text{BOD mg/lt.} = \frac{(D_1 - D_2) - (B_1 - B_2) \times 100}{\%\ \text{sample}}$$

BOD applying seed correction :

$$\text{BOD mg/lt.} = \frac{(D_1 - D_2) - (C_1 - C_2) \times 100}{\%\ \text{sample}}$$

## Volume correction

This correction is applied in case of low BOD sample were larger volume is incubated. In such cases the fraction of dilution water used to dilute the sample is much less than 100% & hence depletion due to blank for the added fraction only should be taken into account & not the 100% in normal cases. The following modified equations is to be used for applying this correction

BOD without applying seed correction:-

$$\text{BOD mg/lt.} = \frac{(D_1 - D_2) - (B_1 - B_2) \infty 100}{\%\ \text{sample}}$$

Where

A = fraction of dilution water added & it given as

$$A = \frac{(100) - (\%\ \text{sample added})}{100}$$

## EXPRESSION OF RESULTS

BOD is expressed as mg/lt, 5 days °C.

## PRECAUTIONS

1. The trend of adding nutrients should be phosphate buffer, magnesium sulphate, calcium chloride & ferric chloride.
2. After aerating dilution water stabilize it at 20°C for atleast one day.
3. The pH of the sample should be in range 6.5 to 8.5.
4. While siphoning the dilution water & filling BOD bottles avoid entrapping of bubbles.
5. Wherever necessary apply seed & volume corrections.

6. Add water on the neck of the bottles during incubation period to ensure continuous & proper water seeing

7. Requisite treatment should be given to the samples containing interfering radicals like iron, residual chlorine etc.

# Chapter -43

# MEASUREMENT OF CHEMICAL OXYGEN DEMAND (COD) IN ENVIRONMENTAL SAMPLES

## GENERAL

The chemical oxygen demand determines the amount of oxygen required for chemical oxidation of organic matter using a strong chemical oxidant, such as potassium dichromate under reflux conditions. The test is widely used to determine :

1. The degree of pollution in water bodies & their self purification capacity.
2. Efficiency of treatment plants.
3. Pollution loads. and,
4. Provides rough idea of BOD which can be used for BOD estimation.

The limitation of the test lies in its inability to differentiate between the biologically oxidizable & biologically inert material.

COD determination has an advantage over BOD test in that the results can be obtained in less than 5 hrs whereas BOD determination requires 5 days. Further, the test is relatively easy & precised. Also there are not many interferences as in the case of BOD.

## PRINCIPLE

Most of the organic matter are destroyed when boiled with a mixture of potassium dichromate & sulphuric acid producing carbon di oxide & water. A sample is refluxed with a known amount of potassium dichromate is titrated against ferrous ammonium sulphate. The amount of dichromate consumed is proportional to the oxygen required to oxidize the oxidizable organic matter.

## INTERFERENCES

Straight chain aliphatic compounds, aromatic hydrocarbons, fatty acids, chlorides, nitrite & iron are the main interfering radicals. The interference caused by chlorides can be eliminated by the addition of mercuric sulphate to the sample prior to addition of other reagents. About 0.48 of $HgSO_4$ is adequate to complex 40 mg $Cl^{-1}$ ions in the form of poorly ionized HgCl.

Nitrite exerts a COD of 1.1 mg/mg N. 120 mg of sulphamic acid is added to $K_2Cr_2O_7$ solution to avoid interference caused by $NO_2$. When 20 ml. sample & 10 ml. dichromate is taken, this can take care of concentration upto 6 mg/lt $NO_2^-$-N.

Silver sulphate is added to conc. $H_2SO_4$ (22g/4kg acid) as a catalyst. This accelerates the oxidation of straight chain aliphatic & aromatic compounds. For complete & better oxidation of organic matter it is necessary to see that the final concentration of $H_2SO_4$ is 50%.

## APPARATUS

### Reflux apparatus

Consisting of a flat bottom, 250-500 ml. capacity flask with ground glass joint & a condenser with 24-40 joint.

### Hot plate

It is convenient to have a bigger hot plate on which a series of refluxing sets is connected to have a single water circulation to all. This eliminates the botheration of having separate blank for each sample.

## REAGENTS

### Standard Potassium dichromate (0.25 N)

Dissolved 12.259 g potassium dichromate previously dried at 103°C for 24 hrs. in distilled water. Add about 120 mg sulphamic acid to this. dilute to 1 lt.

**Sulphuric acid**

Add 22 g silver sulphate to 9 lb. (4 kg) Conc. $H_2SO_4$ bottle. Keep overnight for dissolution. Shake well after dissolution.

**Standard ferrous ammonium sulphate (1 N)**

Dissolve 39 g Fe $(NH_4)_2$ $(SO_4)_2$ $6H_2O$ in distilled water. Add 20 ml. conc. $H_2SO_4$ & cool then dilute to 1 lt. Standardize this solution daily against the standard $K_2Cr_2O_7$.

## STANDARDIZATION

Dilute 10.00 ml. standard $K_2$ $Cr_2O_7$ solution to about 100 ml. Add 30 ml. Conc. $H_2SO_4$ & cool. Add 3-4 drops of ferroin indicator & titrate with ferrous ammonium sulphate till the colour changes to wine red.

$$\text{Normality of ferrous ammonium sulphate} = \frac{10 \times 0.25}{\text{ml. Fe }(NH_4)_2\ (SO_4)_2}$$

## FERROIN INDICATOR

Dissolve 1.485 g phenanthroline mono-hydrate & 635 mg $FeSO_4$ $7H_2O$ in water & dilute to 100 ml. This indicator solution may be purchased already prepared.

## *MERCURIC SULPHATE* : ($HgSO_4$)

## PROCEDURE

Place 0.4g $HgSO_4$ in reflux flask. Add 20 ml. or an aliquet of sample diluted to 20 ml. with distilled water. Add 10 ml. standard $K_2Cr_2O_7$ followed by slowly 30 ml. sulphuric acid which already contains silver sulphate. This slow addition alongwith swirling prevents loss of volatile material such as fatty acid in the sample. Mix well if the colour turns green either take fresh sample with smaller aliquot or add more dichromate & acid. Final concentration of conc. $H_2SO_4$ should be always 50%.

Connect the flask to condenser & reflux for 2 hrs. cool & wash down with condensers with small quantity of distilled water. Remove the flask & add about 80 ml. distilled water. Cool and titrate against standard ferrous ammonium sulphate using ferroin as indicator. Colour changes sharply from green blue to wine red reflux a reagent blank under identical condition preferably simultaneously with sample.

## ALTERNATE PROCEDURE FOR LOW-COD SAMPLES (below 50 mg/l.)

The method is same as for the high COD samples expect in this case the concentrations of $K_2Cr_2O_7$ & ferrous ammonium sulphate is reduced i.e. $K_2Cr_2O_7$ (0.05N) & ferrous ammonium sulphate (0.0025 N).

Exercise extreme care with this procedure because even a trace of organic matter in the glassware or the atmosphere may cause a gross error. COD values upto 160 mg/lt. can be estimated by this procedure.

## CALCULATIONS

$$\text{mg/l. COD} = \frac{(a - b)\,N \times 8000}{\text{ml. sample}}$$

Where,

a = ml. Fe $(NH_4)_2$ $(SO_4)_2$ required for blank

b = ml. Fe $(NH_4)_2$ $(SO_4)_2$ required for sample

N = normality of Fe $(NH_4)_2$ $(SO_4)_2$

## EXPRESSION OF RESULTS

Results are expressed as mg/lt of $O_2$. Report to the nearest whole number.

## PRECISION & ACCURACY

It depend upon the COD value. For the sample having COD higher than 400 mg/lt. precision upto 2% is expected from a good analyst. As the COD value goes on decreasing. Precision also become poorer & poorer i.e. % goes on increasing.

Precision for the low COD samples may be improved by using alternate method where diluted reagents are used. The accuracy of this method is till not very clear. For most organic compounds the oxidation is 95 to 100% of theoretical values.

## PRECAUTIONS

1. The strength of sulphuric acid in final solution be 18 N.

2. The trend of making mixture should be $HgSO_4$ sample (swirl) $K_2Cr_2O_7$ & Conc. $H_2SO_4$ (slowly with swirling).
3. As far as possible reflux blank & samples simultaneously.
4. After refluxing, cool, use required amount of distilled water for washing the condenser. Cool & then titrate.

# Chapter -43

# MEASUREMENT OF HARDNESS IN ENVIROMNETAL SAMPLES

## GENERAL

Hardness in water is mainly due to calcium & magnesium ions. Though some other polyvalent ions such as aluminium, iron, magnanese, strontium & zinc also contribute to hardness. In natural waters, it is defined as sum of the calcium & magnesium ions expressed as calcium carbonate.

Hardness in water posses problems of scale formation in boilers. It is also objectionable for laundry & domestic purposes, since it consumes a larger amount of soap.

Hardness may range from zero to hundreds of milligrams per litre in terms of calcium carbonate, depending on the source of water.

## EDTA TITRAMETRIC METHOD

## PRINCIPLE

As a pH 10 ± 1 EDTA forms a soluble chelated complex with calcium & magnesium. Calcium & complexed magnesium irons also forms a less stable complex with Riochrome Black I indicator. Thus When EDTA is added to a flask, metal indicator complex breaks to form a more stable metal - EDTA complex & when all the metal is complexed to EDTA indicator sows a sharp change in colour from wine red to blue.

Magnesium iron must be present to yield a satisfactory end point in the titration. A small amount of complexometrically neutral magnesium salt of EDTA is therefore added to the buffer.

Thus, this step automatically introduces sufficient magnesium and at the same time obviates a black correction.

## INTERFERENCE

Some metal ions interfere in the end point detection which can be overcome by addition of inhibitors. Usually this interference is unlikely in surface water.

## BUFFER SOLUTION

1. Dissolve 16.9 g ammonium chloride ($NH_4Cl$) in 143 ml. concentrated ammonium hydroxide ($NH_4OH$), add 1.25g of magnesium salt of EDTA & dilute to 250 ml. with distilled water.
2. If the magnesium salt of EDTA is not available, dissolve 1.79 disodium salt of ethylenelimine tetracetic acid dihydrate (analytical reagent grade) & 780 mg $Mg\ SO_4 \cdot 7\ H_2O$ or 644 mg $MgCl_2$ in 50 ml. distilled water. Add this solution to 16.9 g $NH_4Cl$ & 143 ml. conc. $NH_4OH$ with mixing & dilute to 250 ml. distilled water.

Store the solution 1 or 2 hrs. in a plastic container. Do not store more than one month's supply in a frequently opened container.

## INDICATOR

Mix 0.5 g of Eriochrome black T indicator with 100 g of solid sodium chloride. Grind together in a mortar to obtain a homogenous mixture of 40 to 50 mesh size. Use about 0.2 to 0.4 g of ground mixture for each titration.

## STANDARD EDTA TITRANT (0.01 N)

Dissolve 3.723g analytical grade reagent of disodium salt of EDTA in distilled water & dilute to 1 litre. Standardize against standard calcium. This solution is about 1.00 mg $CaCO_3$/ml.

## STANDARD CALCIUM SOLUTION (0.01 M)

Weight 1.000 g anhydrous calcium carbonate into a 500 ml. conical flask. Place a funnel on the neck of the flask & add slowly

1m ml. + HCl until all the $CaCO_3$ has dissolved. Add 200 ml. distilled water & boil for a few minutes to expel out all $CO_2$ Cool and a few drops of methyl red indicator & adjust to the intermediate orange colour by adding 3 N NaOH or 1 + 1 HCl, as required. Transfer quantitatively to a 1 lt. Volumetric flask & make up the volume to the mark. This standard solution is equivalent to 1.00 mg $CaCO_3$/ml.

## PROCEDURE

Take 100 ml. sample or smaller appropriate portion diluted to 100 ml. So that hardness is in range 10-20 mg, in a porcelain dish or conical flask. Add 3 ml. buffer solution followed by about 0.2-0.4 g of solid indicator. Titrate immediately but slowly with continuous stirring, until the last reddish tinge disappears & blue colour is observed.

Reagent blank may be use for comparison. Reagent blank is titrated in a similar way as for the sample. If the hardness is too low, larger volume may be used & also increase the other reagents & indicator proportionately. Blank may be carried out under identical conditions. Standardize EDTA titrant with 25 ml. calcium standard in a similar way as described for the sample.

## CALCULATIONS

$$\text{Hardness as mg/lt. } CaCO_3 = \frac{(V_1 - V_2)}{\text{ml. sample}} \times 1000 \times \frac{25}{V}$$

Where

$V_1$ = Vol. of EDTA required for sample

$V_2$ = Vol. of EDTA required for sample

$V_3$ = Vol. of EDTA required for 25 ml. standard calcium.

## EXPRESSION OF RESULTS

Hardness is expressed as mg/lt. $CaCO_3$. Round off the results to nearest whole number.

## PRECISION & ACCURACY

The precision & accuracy depends on hardness concentration. When the concentration is more than 25 mg/lt, precision & accuracy of about ± 2 to 4% can be achieved.

## PRECAUTIONS

1. Carry out the titration within 5 minutes to minimise the tendency toward $CaCO_3$ precipitation.
2. The final volume of sample & blank should be kept constant usually 100 ml. is better.
3. Use deioniszed or double distilled water for dilutions, blank estimation etc.
4. Use reagent blank for colour comparison.
5. Add same quantity of indicator for sample & blank.
6. Adjust volume of sample such that hardness is in range 5 to 20 mg.

## Chapter -44

# MEASUREMENT OF SULFATE IN ENVIRONMENTAL SAMPLES

## GENERAL

Sulfate is widely distributed in nature & may be present in natural waters in concentrations ranging from a few to several thousand milligram per litre. Mine drainage wastes may contribute high Sulfate by virtue of pyrite oxidation. Sodium & magnesium sulfate exert cathetic action & hence its concentration above 250 mg/l in potable water is objectionable. Sulfates cause a problem of scaling in industrial water supplies & problem of odour & corrosion in wastewater treatment due to its reduction to $H_2S$.

Though the gravimetric method is most accurate for a Sulfate concentration above 10 mg/lt., natural waters are likely to have low concentrations & hence turbudimetric method which is more rapid & also is more or less equally accurate as the farmer method, is suggested. Turbidimetric method may be applied to the concentrations upto 60 mg/lt. When the concentration exceeds 60 mg/lt. it may be diluted accordingly.

## PRINCIPLE

Sulfate ions are precipitated in a hydrochloric acid medium with barium chloride so as to form a uniform suspension of $BaSO_4$ crystals. The absorbance of the suspension is measured by a photometer & the Sulfate concentration is determined by comparisons of the reading with a standard curve.

## INTERFERENCE

Colour & suspended matter interfere in determination. Some suspended matter may be removed by filtration. If both are small in comparison to sulfate on concentration, this interference is corrected by running blanks in which $BaCl_2$ is eliminated. Silicates in excess of 500 mg/lt. will interfere.

There are no ions other than Sulfate in normal waters that will form insoluble compounds with barium under strongly acid conditions.

## MINIMUM DETECTABLE CONCENTRATION

Approximately 1 mg/lt Sulfate.

## APPARATUS

### a) Magnetic stirrer

It is better to use a magnetic stirrer to obtain constants stirring rate & time (usually 1 min. is sufficient).

### b) Photometer

One of the following required :

1. Nephalometer.
2. Spectrophotometer (at 420 nm) and,
3. Filter photometer (at 420 nm, violet blue filter).

### c) Stop watch

For measuring time lapsed for completion of chemical raction.

## REAGENTS

### Conditioning reagent

Mix 50 ml. glycerol with a solution containing 30 ml. conc. HCl, 300 ml. distilled water, 100 ml. 95% ethyl alcohol & 75 g NaCl.

### Barium chloride ($BaCl_2$)

It should be crystalline, 20-30 mesh.

### Standard Sulfate solution

a) Dissolve 1.4798 g anhydrous $Na_2SO_4$ in distilled water & dilute to 1000 ml.

Dilute 100 ml. of solution (a) to 1 lt. This solution is 1 ml. = 100 µg $SO_4.^{-2}$

## PROCEDURE

Take 100 ml. conical flask. Add exactly 5.0 ml. conditioning reagent & mix it on the magnetic stirrer. Add a spoonful of $BaCl_2$ crystals while stirring. Continue stirring for exactly 1 minute at a constant speed.

Immediately after the stirring is over, four some sample into the absorption cell of the photometer & measure the turbidity after every 30 sec. for 4 minutes at 420 nm (violet blue filter). Maximum turbidity usually occur within 2 min. and the reading remains constant thereafter for 3 to 10 min.

Prepare standard curve by carrying standard Sulfate solution through entire procedure. Space standards at 1 mg/lt. Increment in the range 0-40 mg/lt.

Run reagent blank under identical conditions, it is convenient to adjust the photometer (either 100% transmission or zero absorbance) with this blank before taking the sample readings.

To account for the interference due to the colour & turbidity of the sample, run sample blank in which only $BaCl_2$ is withheld. For a particular sample, either this may be used to adjust the instrument or the absorbance due to this is deducted from the sample absorbance. Use same amount of sample for the sample blank & sample determination.

## CALCULATION

$$\text{Sulfate in mg/lt} = \frac{A \times 1000}{\text{ml. sample}}$$

Where

A = conc. of Sulfate ions in mg., obtained from the calibration curve for the corresponding sample absorbance.

## EXPRESSION OF RESULTS

Results are expressed as mg/lt. Sulfate.

## PRECAUTION

1. Time & rate of stirring should be kept constant throughout the determination i.e. for standards, blank & samples.

2. Final volume should be kept constant.
3. Make all turbidity measurements after a constant selected time, when the turbidity is maximum.
4. Wherever required use sample blank to account for the interference due to turbidity & colour of the sample. Suspended matter may be removed by filtration,
5. Method holds good up to concentration 60 mg/lt. For higher concentration, dilute proportionately.

## ARGENTOMETRIC METHOD

## PRINCIPLE

Chloride is determined in a natural or slightly alkaline solution by titration with standard silver nitrate, using potassium chromate as indicator. Silver chloride is precipitated quantitatively before red silver chromate is formed.

## INTERFERENCES

Bromide, iodide cyanide are measured as equivalent to chloride ion. Sulphide, thiosulphate, cyanide thiocyanate & sulphites interfere. This may be oxidized to non-interfering substances as described below :

Take a known volume of sample, add 25 ml. of hydrogen peroxide 3% & boil for 15 minutes. Add further 10 ml. of hydrogen peroxide & boil for 5 minutes. Test by spotting out for the absence of thiocyanate with ferric chloride. Formation of yellowish orange spot indicates presence of thiocyanate. Repeat the heating with hydrogen peroxide till it becomes free from thiocyanate. Orthophosphate in excess of 25 mg/lt. interferes by precipitation as silver phosphate. Iron in excess of 10 mg/lt. interfere by masking the end point.

## REAGENTS

### Silver nitrate standard solution, (0.028 N)

Dissolve 4.791 gms. Silver nitrate in distilled water & dilute it to 1000 ml. Store in brown glass bottle. So,

1.0 = 1.0 mg $Cl^-$

Standardization & titration must be carried out without any dilution.

### Potassium chromate (Indicator) solution

Dissolve 50 g $K_2CrO_4$ in a little distilled water. Add silver nitrate solution dropwise until definite red precipitation starts. Let it stand for 12 hrs. Filter & dilute to 1 lt. with distilled water.

Use chloride free i.e. redistilled or deionized distilled water for preparation of all above reagents. Same water should be used for sample dilution also (wherever required). Avoid dilution as far as possible or apply blank correction.

## PROCEDURE

Take 100 ml. sample or a suitable portion diluted to 100 ml. into a porcelain dish or conical flask or beaker on a white surface.

Tritrate samples in the pH range 7 to 10 directly. Adjust samples if not in this range with $H_2SO_4$ or NaOH solution. Add 1 m,. potassium chromate solution and titrate with silver nitrate solution with constant stirring until only the slightest reddish coloration persists. Be consistent in end point recognition.

It is suggested to use reagent blank as comparator. If the volume of the titrant exceeds 25 ml. then it is better to repeat the method with smaller portion of the sample.

Standardize the silver nitrate titrant standard sodium chloride. Carry out the blank determination to account for the chloride present in any of the reagents & for the solubility of silver chromate. A blank of 0.2 to 0.4 is usual for this method

## CALCULATION

Chloride as

$$Cl- = \frac{(V_1 - V_2)}{\text{Vol. of sample in ml}} \times \frac{25}{V_3}$$

Where,

$V_1$ = volume of silver nitrate required for sample.

$V_2$ = volume of silver nitrate required for reagent blank and,

$V_3$ = Volume of silver nitrate required for standard sodium chloride.

## EXPRESSION OF RESULTS

Results are expressed as mg/lt. $Cl^{-1}$. The results should be rounded off as follows:-

Concentration results mg/lt. less than 20 but above to 10, record to nearest mg/lt 0-1 to whole number

## PRECISION & ACCURACY

Precision & Accuracy of about ± 2 to 4% is expected from a good analyst, except in the sample, where chloride content is less than 50 mg/lt.

## PRECAUTIONS

1. Use white opaque background for titration, it is suggested to carry out titration in white porcelain dish.
2. The final volume of titrant exceeds 25 ml., it is suggested to make a smaller quantity of the sample. Also, the volume of titrant should not be too small in such cases volume of sample may be increased. It is better to adjust the volume of the sample such that titrant value is in range of 5-20 ml.
3. Sample is coloured which may interfere with the chromator end point in such cases mercurimetric methcd should be used.
4. Be consistent in end point recognition. Reagent blank may be used for comparison.
5. Use deionised distilled water for preparation of solution as well as any dilutions.

## MERCURIC NITRATE METHOD

## PRINCIPLE

Chloride ions from a highly stable complex with mercuric ions, thus they may be titrated with a standard mercuric nitrate. The end point is detected by diphenylcarbazone indicator in the pH range 2.9 to 3.4, which forms a blue violent compel with an excess of mercuric ions.

## INTERFERENCE

Iodide, bromides, cyanides & thiocyanides are measured as equivalent of chlorides sulphide, thiosulphate, sulphite, cyanide & thiosulphate interfere seriously with the determination. They may be oxidized to non interfering substance as described in argentometric method.

If the sample is highly coloured or turbid to allow the end point to be readily seen this interference may be reduced by the aluminium hydroxide suspension treatment.

Add 3 ml. aluminium hydroxide suspension to the measured quantity of sample. Stir thoroughly, let is settle for few minutes & filter. Wash the precipitate with 10-15 ml. distilled water. Also collect washing with the filtrate & continue as described earlier.

Chromate, ferric, zinc, copper, manganeous & sulphite ions interfere in concentration above 10 mg/lt., aluminium, lead & nickel in concentrations above 100 mg/lt. In such cases these titrimetric methods are not suitable.

Phosphates do not interfere.

## REAGENTS

### Mercuric nitrate solution (0.0282 N)

Dissolve 5.048 g mercuric nitrate ($Hg(NO_3)_2\ 2H_2O$) in 50 ml. water containing 0.5 ml. conc. nitric acid dilute to 1 lt. & filter if necessary.

### Nitric acid 0.05 N

Dilute 3.2 ml. conc. nitric acid (density 1.42 g/ml) to 1 lt.

### Sodium hydroxide 0.05 N

Dissolve 2 g sodium hydroxide in distilled water & dilute to 1 lt.

## PROCEDURE

Take 100 ml. or smaller portion so that the chloride content is less than 20 mg in a proclain dish. Add 5-10 drops of indicator. If a blue, violent or red colour develops, add 0.05 N nitric acid drop by drop until it becomes yellow & then add exactly 1 ml. in excess.

Titrate the solution with standard mercuric nitrate to the point where the first tinge of blue-violent colour appears & does not disappear on stirring.

Prior warning that the end point near is given by the appearance of orange colour. Determine the reagent blank by titrating 100 ml. of distilled water by the same procedure.

## CALCULATION

$$\text{Chlorides as } Cl^- = \frac{(V_1 - V_2) \times 1000}{\text{Volume of sample (ml.)}} \times \frac{25}{V_3}$$

Where,

$V_1$ = volume of mercuric nitrate required for sample.

$V_2$ = volume of mercuric nitrate required for reagent blank and,

$V_3$ = volume of mercuric nitrate required for 25 ml. of standard sodium chloride.

## EXPRESSION OF RESULTS

Record the results as mg/lt. $Cl^-$

The results should be rounded off as follows:-

Conc. range (mg/lt) - 1-20, record to the nearest (mg/lt) 0.1 & 20 onwards to the nearest whole number.

## PRECISION & ACCURACY

The precision & accuracy of about ± 2 to 4% is expected from a good-analyst. This also depends on the quantity of chloride present & on the pretreatment necessary.

## PRECAUTIONS

1. Titration should be carried out on a opaque white base for which white porcelain dish may be used.
2. To final volume should be kept constant say 100 ml.
3. The volume of sample should be selected such that it should not contain more than 20 mg of chloride. On the other hand, volume of sample may be increased if $Cl^-$ content is very small.
4. Deionized (chloride free) distilled water should be used for preparation of solutions as well as for dilution etc.

Chapter -46

# MEASUREMENT OF ALKALINITY IN ENVIRONMENTAL SAMPLES

## GENERAL

The alkalinity of a water is its quantitative capacity to neutralize a strong acid to a designated pH. Since the alkalinity of many surface waters is primarily a function of carbonate, bicarbonate and hydroxide ions, the alkalinity is taken as an indication of the concentration of these constituents. The measured values may also include contributions from borates, phosphates or silicates if these are present.

The alkalinity of many natural and treated waters is due only to bicarbonates of calcium and magnesium. The pH of these waters does not exceed 8.3. Their total alkalinity is practically identical with their carbonate hardness and such samples do not have phenolphthalein alkalinity.

Waters having a pH above 8.3 contain besides bicarbonates, normal carbonates and possibly hydroxides also. The alkalinity fraction equivalent to the amount of acid required to lower the pH of the sample to 9.3 is called phenolphthalein alkalinity. This fraction is contributed by half of the carbonate and by the hydroxide, if present.

The amount of acid required to lower down the pH to 4.5 determines the total alkalinity or methyl orange alkalinity.

Alkalinity values provide guidance in applying proper losses of chemicals in water and wastewater treatment process particularly in coagulation, softening and operational control of an aerobic digestion. The alkalinity in excess of alkaline earth concentrations is significant in determining the suitability of a water in irrigation.

## METHODS

The alkalinity is determined by titration of the sample with a standard solution of a strong mineral acid. Titration is carried out either using indicator or a calibrated pH meter.

The simple and rapid visual method using an indicator is satisfactory for control and routine applications. Electrometric method is used for accurate determination. Electrometric method is used for accurate determination. It is also free from interference such as colour, turbidity or suspended matters.

Low alkalinities (below 20 mg/l $CaCO_3$) are also best determined by a eletrcometric titration. Thus, for natural or surface waters were low alkalinity is expected electrometric method is preferred over the visual indicator method.

Wherever possible, the titration should be varied out of the point of sampling. Otherwise, the sampling bottle should be completely filled and the alkalinity is determined within 24 hrs.

## INTERFERENCES

Coloured, turbidity and suspended matter may interfere with the visual titration method by marking the colour change of the indicator. Turbidity and suspended matter can be eliminated by simple filtration.

Electrometric method is free from all above interferences.

## ELECTROMETRIC TITRATION METHOD

### Apparatus

Any pH meter utilizing a glass electrode may be used. It must be standardized according to the directions supplied by manufacturer. It is convenient to use a burette with a bent jet, so that the burette can be used in an upright position.

## REAGENTS

### Carbon dioxide free distilled water

It must be used for the preparation of all standard solutions. If the distilled water has a pH lower than 6.0, it should be freshly boiled for 15 minutes and cooled to room temperature.

### Sodium carbonate solution (0.05 N)

Dilute 3.1 ml. conc. Sulphuric acid (or 10 ml. conc. hydrochloric acid) to 1 lt. Standardize against 40.0 ml. 0.05 N $Na_2CO_3$ solution with about 60 ml. water in a 250 ml. beaker of titrating electrometrically to the pH inflection point. This can also be standardized using methyl orange indicator.

### Sulphuric acid (or hydrochloric acid) (0.02 N)

Dilute 200 ml. of 0.1 N standard acid to 1 litre with carbon dioxide free distilled water. Standardize using 10 ml. of 0.05 N $Na_2CO_3$ by electrometric or visual titration method.

Calculate the normality from equation given below :-

$$\text{Normality of acid, N} = \frac{X.\ Y}{5300.Z}$$

Where,

X = $Na_2CO_3$ dissolved in 1 lt. (g).

Y = $Na_2CO_3$ taken for titration (ml) and,

Z = Standard acid consumed in titration (ml.).

When concentration of sulphuric acid is adjusted exactly to 0.1000 and 0.0200 N, then:-

1 ml. of 0.1000 N solution = 5.00 mg $CaCO_3$

1 ml. of 0.2000 N solution = 1.00 mg CaCO3

## PROCEDURE

Take 100 ml. sample in a 250 ml. beaker Place it on a magnetic stirrer. Dip the electrodes into the solution and adjust the top of the burette on the surface of the sample.

Titrate the standard acid to pH 8.3. Record the volume required, this corresponds to the phenolphthalein alkalinity

Continue the titration to a pH 4.5. The total titration value represents the total alkalinity.

## VISUAL TITRATION METHOD

### Reagents

***Carbon di-oxide free distilled water***

Same as described for the electrometric method.

***Sodium Carbonate 0.05 N***

Same as described earlier.

**Phenolkpthalein indicator**

Dissolve 0.5 g of phenolphthalein in 50 ml. of 95% ethanol and add 50 ml. of distilled water. Add to it dilute (e.g. 0.01 N or 0.05N) carbon di oxide free solution of sodium hydroxide slowly drop by drop till a faint pink colour persists.

**Methyl orange Indicator**

Dissolve 0.5 g of methyl orange in 100 ml. water.

## PROCEDURE

Add 3-4 drops of phenolphthalein indicator to 100 ml. of water in a porcelain dish or a conical flask over a white surface. If no colour is produced, the phenolphthalein alkalinity is zero.

If the sample turns to pink titrate with standard acid till pink colour just disappears. In both cases continue the determination using the sample to which phenolphthalein has been added.

Add a few drops of methyl orange indicator, if orange colour is observed without adding any acid, the total alkaninity is zero. If the sample turns yellow, titrate with standard acid until a first appearance of orange colour is noticed.

## CALCULATIONS

$$\text{Phenolphthalein alkalinity (pH 8.3, in mg/lt. as } CaCO_3) = \text{A.N. } 50000$$

$$\text{Total alkalinity (pH 4.5. in mg/lt. as } CaCO_3) = \text{B.N. } 50000$$

Where,

A = is the ml. of standard acid required to reach the phenolphthalein end point or pH 8.3.

B = is the ml. of standard acid required to reach the methyl orange end point or pH 4.5. and,

N = is the normality of acid used for the titration.

## EXPRESSION OF RESULTS

Results are expressed as mg/lt. of $CaCO_3$. For conversion purposes, it may be recalled that 50 mg/lt. as $CaCO_3$ are equivalent to 1 mg/lt.

## PRECISION AND ACCURACY

No general statement can be made about precision because of the great variation in sample characteristics.

For total alkalinity higher than 50 mg/lt., precision and accuracy of electrometric titration are estimated around 1% while for the lower concentration, it may be around 2% to 3%.

The precision and accuracy of visual titration is estimated at 2% for the alkalinity above 50 mg/lt. as $CaCO_3$ and 2% to 10% for alkalinity between 500 mg/lt. and mg/lt. as $CaCO_3$.

## PRECAUTIONS

1. Do not dilute, concentrate or alter the sample in any way. As far as possible also avoid filtration.
2. Use only carbon dioxide free distilled water for all purpose such as preparation of reagents etc.
3. If possible, titration should be carried out on the spot. If not, the sampling bottle should be filled completely and the alkalinity

is determined in shortest possible time. In any case, not later than 24 hrs.

4. Use magnetic stirrer for stirring purposes.

5. Adjust the volume of sample and concentration of titrant such that the volume of titrant required is appropriate i.e. neither too low nor too high.

6. While preparing 0.1 N $NH_2SO_4$, add conc. $H_2SO_4$ to water (not water to $H_2SO_4$) cool then make up the volume.

Chapter -47

# AUTOMATIC SAMPLING & MONITORING TECHNIQUES FOR ENVIRONMENTAL SAMPLES

## GENERAL

The use of automatic monitoring instrumentation for air & water is increase and in many cases it is replacing manual analytical techniques, because it has following advantages :

1. Continuous recording is possible (as & when it is found essential).
2. Comparatively rapid analysis of samples & subsequently equally quick processing of collected data.
3. ease to operate.
4. sufficient accuracy. and,
5. least man power requirement.

The system used for measuring air & water pollution parameters generally include following steps, although not essentially in the same sequence :

1. Sampling.
2. Sample Preservation.

3. Sample Pre-treatment & pre concentration.
4. Data acquisition & reduction.

There are 3 types of instruments :

1. ***Laboratory or manual instruments***

These are generally used in the laboratory and need manual operation. It includes spectro-photometer, chromatographs etc. and follows all the step are required.

2. ***Semi-automatic instruments***

These measure the parameter on interrupted or sampled basis & analysis on this discrete samples. The process is automatically repeated on a regular basis. Automated GC is the one of the example of this type instrument. These are also known as batch sampling type instrument.

3. ***Continuous sensor type instruments***

These measure the constituent or parameters or an uninterrupted basis.

Again they are of two types :

a. Those requiring Pre-treatment of sample or addition of chemicals like buffer etc. Ion selection electrodes for some specific ions in water stream & conductometer for atmospheric $SO_2$ conc. measurement are the typical examples.

b. Those requiring no addition of chemicals or pre-treatment of samples. pH, temperature, DO, conductivity, turbidity etc. in water sample & $NO_X$ & $O_3$ analyser for air pollution analysis are the typical example of this type instruments.

## MEASUREMENT TECHNIQUES

Although wet chemicals procedures filled the early needs of measurement of environmental pollution parameters, especially of air, pollution, these are becoming obsolete.

Monitors using ion selective electrodes, atomic absorption, anodic stripping voltametry, Gas chromatography, mass spectrometry, flame photometer, etc. are second generation instruments are generally in use for environmental pollution control studies.

## WATER POLLUTION MONITORS

Table 47.1 gives the complete list of instrumental techniques used for water monitoring.

**Table 47.1 : Instrumental techniques used for water pollution parameters**

| *Type* | *Parameters* | *Techniques used* |
|---|---|---|
| *1* | *2* | *3* |
| Physical & Chemical | pH-conductivity Hardness | Electrodes-bridge/ Ion selective electrodes & colorimetry |
| | Salinity | Conductivity (AS) |
| | Temperature | Thermocouples |
| | Turbidity | Nephelometry |
| | Flow & level | Float & magnetic tape |
| Biochemical & microbiological | BOD | Respirometer & D.O. Probe |
| | C.O.D. | Colorimetric |
| | D.O. | Membrane probe |
| | Total Organic Carbon | TOC analyser |
| | Bacteria/Viruses | Colony counter |
| Dissolved gases | Oxygen | D.O. Probe |
| | Chlorine | Colorimetric & electro- chemical |
| | $H_2S$ | Colorimetric/ Iodometric |
| Metals | All metals | AAS |
| | Na, K, Mg, Li | Flame Emission |
| | Pb, Fe, Cu, Na, K | Ion selective electrode |

| | 1 | 2 | 3 |
|---|---|---|---|
| | | Lu, As, An, Cu, Hg | X-ray fluorescence |
| | | Hg | Neutron activation |
| 5. | Halides & Cyanides | $F^-$, $Cl^-$, $Br^-$, $I^-$, $CN^-$, | Ion selective electrode |
| 6. | Ammonia, N, P, S | NH3, $NO_2^-$, $NO_3^-$, P, S, $SO_4$, etc. | Ion selective electrode |
| 7. | Trace toxic compounds | Pesticide, insecticide herbicides, | Gas chromatography |
| | | Chlorinated HC, Petrochemicals | GC-MS |
| | | Phenols, Oil & grease | TLC |

## MODERN WATER MONITORING SET UP

A typical water or waste water treatment plant, as expected to monitoring following parameters :

### 1. Dissolved oxygen (D.O.)

It gives in general the degree of pollution caused by organic compounds, undergoing biological degradation & is a measure for the self-cleaning action of the water. Generally membrane probes are used for D.O. monitoring.

### 2. Flow & level

This is very important in calculating actual pollution concentration & predictions of down stream water quality, float type level sensor & magnetic type flow sensors are used for the measurement.

### 3. pH

One of the fundamental parameters for water quality measurement. Generally membrane probes are used for this.

**4. Redox**

Useful for evaluating the ratio between reducing & oxidising agents. Generally metallic electrodes are used for this measurement.

**5. Chlorides & Conductivity**

Useful for assessment of salinity. Man made pollution can be evaluated if natural chloride content is know. Generally conductivity cell is used.

**6. Selected ions**

Very important parameter for down stream monitoring of industrial waste discharge. Generally $NO_3^-$, $SO_4^{-2}$, $PO_4^{-3}$, $NH_3$, CN, toxic metals like Hg, Cu, Zn, Na etc. can be monitored using the specific ion electrodes.

**7. Temperature**

Very important for monitoring the water quality and self purification capacity of an aquatic system.

**8. Meteorological parameters**

Solar radiation, ambient air temperature, wind speed, direction significantly affect the existance, dispersal, dilution of any environmental pollutant.

## INSTRUMENTS GENERALLY USED FOR AIR POLLUTION MONITORING STUDIES

Table 47.2 gives an idea about the general & sophisticated instruments used in air pollution monitoring studies :

**Table 47.2 : Instruments for air pollution analysis**

| *Parameters* | *Instrumental technique* |
|---|---|
| *1* | 2 |
| $SO_2$ | Conductometry |
| $SO_2$, $NO_X$ | Colorimetry |
| $SO_2$, $NO_x$, Oxidants ($O_3$) CO | Colorimetry, amperometry |

| 1 | 2 |
|---|---|
| $SO_2$ | Paper tape ($H_2S$ conversion) |
| CO, $NO_x$, $SO_2$ | Electrochemical cells ( EMF generation |
| CO | Catalytic oxidation |
| $SO_2$, $NO_x$ | Specific ion electrode |
| $NO_X$, $O_3$ | Chemiluminiscence |
| $SO_2$ | FPD, GC |
| CO, $CH_4$, HC | FID, GC |
| CO | NDIR |
| HC, $SO_2$, $H_2S$ | NDIR |
| Odixant | NDUV |
| $SO_2$ | UV fluorescence |
| $SO_2$, $NO_2$, CO | Bioluminiscence |
| $SO_2$, $NO_X$ | Spectroscopy |

Following Table 47.3 gives an idea about various instruments, used for air pollution studies, specially particulate matter analysis.

**Table 47.3 : Instruments for particulate analysis**

| *Parameters* | *Instruments* |
|---|---|
| All metals | AAS |
| Zn, Cd, Cu, Hg | Atomic fluorescence |
| Mostly All metals | X-ray fluorescence |
| Aromatic & chlorinated HC, pesticides, oxidants | GC-MS |
| Heavy Metals - V, Hg | Neutron activation |
| Metals: Cu, Cd, Pb | Anodic Stripping, Voltametry, Polargraphy |

Chapter -48

# MICROPROCESSOR & DIGITAL DISPLAY FOR ANALYTICAL INSTRUMENTS FOR MONITORING OF ENIVORNMENTAL QUALITY

## GENERAL

Modern electronics offer the analysis with analytical instruments having the features & benefits not previously possible and at a price not too exorbitant.

Mostly all the analytical instruments, now can have digital read-out instead of the conventional meter system. A number of instruments manufactured indigenously viz. pH meter, flame photometer, colorimeter, conductivity meter also have this facility.

Number of sophisticated analytical instruments like AAS, GC, XRF etc. now-a-days have one additional facility of microprocessor or microcomputers.

With the help of these facilities suitable with the instrument, the task of the operators is very much simplified & he has very little to do which saves his considerable amount of valuable time.

## ADVANTAGES OF INSTRUMENT WITH DIGITAL READOUT

In the digital instruments, the signal is first converted into digital form (if required) & is then processed by using special electronic circuits integrated circuits performing various complex logic operations like And, OR, NAND, NOR, Triggering, counting, decoding etc. & finally given to special indicating units like 7 segment LED (light emitting diodes).

The signal value is thus displayed directly in digits. Liquid crystal Display (LCD) system is also quite often used. The LED display actually glow which LFCD requires a small amount of external light.

With digital display, measurement becomes very simple without the scope of any personal error. The ambiguity that is present while reading conventional meters because of parallex etc. is completely avoided.

A significant advantage of digital over analogue instrument is that of increased precision, obtained by employing just few additional digits. Furthermore, the digital signals can be amplified indefinitely & stored accurately. also because of On-Off character of digital circuit, the drift & stability problems are inherently negligible.

## MICROPROCESSOR

A microprocessor is a complex LSI (large scale integrated circuit) chip which can be programmed externally to perform controlled logic operations.

A microprocessor forms the heart of a digital processing system & is used as a central processing unit (CPU) in the microcomputer. The functions to be performed by the micro-processor. When used in analytical instrument is determined by the sequence of steps of operations required for the instrumetn & these are stored in its memory much like a conventional digital computer.

The functions carried out by the microprocessor can be altered by changing the programme stored in the memory which makes it most versatile & flexible device.

Microprocessor can be broadly classified into 4 groups :

1. General purpose type and
2. Dedicated type.

3. Controller type and,
4. Calculator type.

They can be classified according to their CPU bit size viz. 4, 8, 12, 16 bits etc.

A microprocessor system usually comprises of the microprocessor unit (MPU) itself, a Random Access Memory (RAM), a Red Only Memory (ROM), a programmable communication interface, a peripheral interface & priority interrupt controller.

## ADVANTAGES & APPLICATIONS OF MICROPROCESSORS

Microprocessor has already found its way into all branches of electronics from entertainment to computer and control and from communication to measurement & instrumentation (electronic calculators, computers, peripherals, process control, data acquisition terminals, electronic instrumentation & controls, medical equipments, automatic & remote control equipments.

The attempts to employ microprocessors in instruments have met with a great success and with intelligence. The whole concept of instrument design has undergone a radical change with the introduction of microprocessors in electronic instrument.

The advantages of microprocessor, particularly in the analytical instrumentation have been given below :

1. It makes the instruments most compact (size reduction) more versatile (capabilities) & more reliable.
2. Operation becomes very simple with entry of instruments functions via numerical keyboard.
3. Automatic zero, automatic curve correction & linerisation & calibration are possible with the help of microprocessor.
4. Measuring range-wider scale-viz.-0-4 absorbance in case of spectrophotometer is possible.
5. Since the microprocessor can make actual calculation programmed by it not use of logarithmic amplifier is necessary and effecting considerable reduction in drift & a strictly flat base line.

6. Display is digital from directly in the measuring unit % transmission, absorbance or concentration is possible.
7. The microprocessor can also control the external data logging device like recorder, printer, teletypewritter etc. coupled with the main instrument.

## ILLUSTRATION OF MICROCOMPUTER FACILITY

### Perkil Elmer AAS PE 372 - Microprocessor controlled

The micro computer with this instrument is used to elect the integration time & scale expansion to calibrate the instrument to read directly in concentration.

Compensating for deviations for linearity and correcting for the calibration curve and to control external data logging device - printer, recorder, teletypewritters etc.

The programme provided with the micro computer is permanently contained in two devices called ROMS (Read only memory). since these are permanently designed into the system, there is no danger of accidentally changing or erasing the programme.

The another component is called RAM (Random Access Memory) is used to handle information that the operator may wish to change from one analysis to the next, such as concentration of standards and integration time.

To enter the desired functions the operator has to merely press the appropriate buttons on the keyboards & the RAM remembers the information for use as necessary.

The calibration has also been very much simplified. When the instrument is first turned ON the microcomputer automatically selects IX scale expansion & 0.5 seconds integration time.

If the operator wants to select different integration time the same is entered on teh numerical keyboard & T button is pressed on the keyboard.

The lamp current, slit width & wavelength is selected as per standard procedures, depending upon the element to be analysed, e.g. f Mn is to be measured in the range 0-3.0 μg/ml. the wavelength is set to 279.5 nm, the spectral slit width to 0.2 nm & the lamp current to 10 MA.

The flame is lighted with push button ignitor. The blank is then aspirated & the lamp and the flame are oligned to get minimum reading when the 'READ' button is pressed.

Now, the auto zero button is pressed to automatically bring this background reading to zero. Then the Mn standard of 1.0 ppm is aspirated & the reading is noted and at the end of the integration time and this value of standard i.e. 1.0 ppm is entered by pressing the button "S1".

The instrumetn automatically sets the value measured to the concentration (1.0 ppm) entered to "S1".

Then the second standard of 3.0 ppm is aspirated & the instrument is allowed to read. At the end of integration time, this concentration value i.e. 3.0 ppm is entered on the keyboard & "S2" button is pressed.

The instrument is now calibrated to read directly in ppm any sample concentration of Mn between 0-3.0 ppm & well correct any deviations from linearity which may exist in the absorbance/ concentration relationship of Mn.

The curve fitting programme of the microcomputer will automatically correct the curves & substantial deviations from linearity.

## MICROPROCESSOR CONTROLLED IR SPECTROPHOTOMETER (PERKIN ELMER PE 590, 281)

The use of microprocessor is more significant in IRS. The use of microprocessors is more significant in IRS. The use of microprocessors alongwith the integrated scan controls has solved many of the problems like recording conditions, scan time, pen response, slit programme etc.

The operator selects proper recording parameters etc. from integrated controls. The microprocessor then automatically optimises the other conditions. Even under high resolution conditions, where the noise might obscure spectral details, the operator by selecting appropriate multiplier improves the signal to noise ratio, because the instrument then automatically changes teh scanning parameters to optimise the recording condition.

Microcomputer can also help I-R analyst by an additional way. Infra-red spectrum of an organic material provides invaluable qualitative & quantitative information. However, the analyst has to spend lot of time particularly when looking for difference between two closely related compounds.

In the microcomputer controlled instruments the procedures are much simplified, since the same microcomputer which controls the instrumetn also provides rapid and unique data reduction when used along with an ordinate data processor.

The ordinate data processor also provides a large digital reagent values of the parameter % T or A etc. by the operator.

For quantitative work the absorbance values can be expanded or contracted to provide direct readings in concentrations.

A unique all peak mode with the help of digital printer, automatically prints out each peak, as it occurs giving the peak number, its exact frequency to 0.1 $cm^{-1}$ and also the ordinate value against in % T or directly in concentration.

A maximum peak mode automatically calculates & holds the maximum peak observed during the scan. When combined with repetitive scan accessory, the maximum peak is printed at the end of each scan & auto zero is activated before the next scan is initiated.

Thus the minute differences between the two closely related compound can be very easily distinguished.

## Chapter -49

# MAINTENANCE & SERVICING OF ELECTRONIC INSTRUMENTS

### GENERAL

With the increasing awareness of ill effects of the environmental pollution, it has become necessary to detect & measure the concentration of various pollutants even to the pictogram & nanogram levels.

More and more sophisticated and extremely costly analytical instruments are required for this purpose. The instruments like GLC, AAS are now increasingly being used in our country also at pollution control institutions, labs, etc. However merely the procurement of these instrument would not serve much purpose if they are not properly operated & maintained.

Thence maintenance and servicing should be given proper attention. Few important aspects of maintenance and servicing of analytical instruments have been discussed below :

### OBJECTIVES

1. To achieve long, uninterrupted and useful operation from the instrument.
2. To minimise actual close down period thereby saving valuable time & money of uses.
3. To provide maximum safety to user.

The above objectives can be achieved by observing some simple rules as given below :

1. Follow correct operation procedures with due precautions &,
2. Provide periodic & routine check ups and preventive maintenance

## INSTRUMENTS & INSTRUMENT PARTS

It is required to protect the costly analystical instruments and their sensitive & valuable components and parts like :

***Instruments***

1. pH meter
2. D.O. analyser
3. Conductivity meter
4. Turbidity meter
5. Colorimeter
6. Spectrophotometer
7. Flame photometer
8. Specific ion electrodes
9. Polarographs
10. Total organic carbon analyser
11. AAS
12. Mercury Analyser
13. GLC
14. GC-MS.

***Parts** : Various types are as follow :*

***a. Mechanical parts***

1. Gears
2. Bearing
3. Pen point
4. Pen drive
5. Chain drive
6. Chart drive
7. Slide Wires
8. Gas cylinder
9. Pressure regulator
10. Metallic diaphragms
11. Meter covers
12. Burners
13. Automisers
14. Capallaries etc.

***b. Electro mechanical parts :***

1. Coils
2. Relays,
3. Motor
4. Clutches
5. Meter
6. Moving Coil - Spot galvanometer
7. Band switches
8. Press button switches
9. Contacts etc.

***c. Optical parts :***

1. Source lamps,
2. Mirror
3. Lenses
4. UV, IR amps,
5. High pressure mercury vapour lamps
6. Mono chromators
7. Prisms
8. Gratings
9. Optical window etc.

***d. Glass parts :***

1. Electrodes
2. Cells
3. Cuvettes
4. Capallaries
5. Tube
6. Stopcocks
7. Ground glass joints

***e. Electrochemical parts :***

Batteries like lead-acid, alkaline, drycell, mercury battery, standard cells, Ni-Cd (rechargeable) etc., specific ion electrode, electrolytes, buffer, stock solutions, standard solutions etc.

***f. Electronic parts :***

Precision & High value resistance, high voltage condensers, capacitors, filters, coupling, bypass etc. RF coils chokes, spl. transformers, electronic, photo cells, photorestors, phototubes, Hollow, chathode lamps, printed circuit boards, transistors, integrated circuits, digital displays, microprocessor etc.

## PROTECTION REQUIRED

It is necessary to protect the instrument & their different parts from improper usage, handling and storage :

1. Improper and extreme operating voltages.
2. Improper & extreme temperature during operation & storage.
3. Dust.
4. Humidity, Water spillage.
5. Corrosive vapours.
6. Friction-wear & tear.
7. Other environmental factors like vibration, shock, noise, explosion etc.
8. Microorganism - fungus, moulds etc.

## IMPROPER OPERATING VOLTAGE

Instruments should be operated on the specified voltage only. Instrumental voltage & plugging are need proper attention, when the same are joint to be installed & operated in a laboratory, the voltage range & colour indication to various line must be noted down. All the instruments must be protected from high voltage fluctuation, for which stabilisers are required.

### Improper temperature

For storage & operation of a particular instrument a particular temperature range is defined. The instrument should be housed in good ventilated rooms and never exposed directly in the sun. At around 25°C, most of the instruments run satisfactorily. Air conditioned rooms are required for sophisticated instruments. Sensitive components, chemicals, reagents may be kept in sealed containers & stored in refrigerator.

### Protection from humidity

Excessive humidity disturbs the performance of most of the instruments & may cause permanent damage to some sensitive electronic parts like electrometer, high resistance contact, photoelectric cells, optical parts like infra-red cells & windows.

Effect of dust & corrosive fumes and fungi is enhanced in the humid conditions. Infrared instruments are to be specially protected from humidity. Hot & humid conditions also shorten the shelf life of dry cells & mercury batteries.

Sensitive parts may be always stored in dry and cool environment- in desiccator etc. Silica get kept in special compartment of electronic instrument viz. high input resistances - 2000 meg ohms - compartment of electrometer & FET amplifiers used for phototubes & photomultiplier tubes of Turbidimeters, spectro-photometers etc. & glass electrodes of pH meters) should be periodically reactivated.

Hot air blower also used to dry out these sensitive compartments & solve the common troubles caused due to humidity.

Relative humidity of 60% or less is recommended for analytical instruments. A dehumidifier coupled to the air conditioner is advantageous for the operation of sophisticated instruments.

## PROTECTION FROM DUST

The instruments should be located in a dust free and clean environments. Air conditioned room with linoleum flooring may be done used for housing the instrument to minimise the dust problem.

Electronic circuits, high resistances, contacts & especially optical parts like mirror, lense, grating are very much affected by dust combined with moisture.

Dust cover must be used for the instruments during their idle period & the instruments should be cleaned periodically.

Different parts require separate treatment :

a. ***Metallic parts***

A vacuum cleaner fitter with a small hair brush tip may be used initially to remove the dust. Metallic parts may be further cleaned by cotton soaked with household detergent of soap solution. Electronic circuits can be simply cleaned by a hair brush.

b. ***Glass part***

Ordinary glass parts may be cleaned by chromic acid and followed by rinsing with distilled water. Household detergent or soap solution may be used to remove oil & grassy stains while organic solvents like alcohol, acetone may be used to remove scaling due to organic material.

c. ***Optical parts*** :

Optical parts like mirrors, grating, monchormators require special care & precaution while handling and cleaning. Scratches & even finger prints should be avoided, as it would affect their normal functioning.

1. Mirrors & lenses may be periodically cleaned using soft tissue or lens paper or air bulb syinges & may be occasionally cleaned by using lint or cotton swab fixed on the tip of a wooded stick soaked with a very mild detergent or alcohol. Xylol is best suited for cleaning microscope optical parts.

2. Grating & monochromator should not be touched or disturbed as far as possible. Only occasionally the settled dust may be cleaned by using air bulb syringe or very soft hair brush.

3. Sample cells & cuvettes should be always kept clean and without finger prints. Mild detergent or organic solvent like alcohal may be used for washing these parts & subsequently rinsing several times by distilled water & finally blotting by tissue paper.

**Effect due to corrosive vapours, scaling, coating etc.**

Fumes & vapours of various chemicals & detergents may affect the instruments & their parts. Some fumes (acidic) are very much corrosive, while some other (organic vapours) may affect optical parts. Mercury vapours are especially harmful to the optical parts like grating, monochromators etc.

Various activities like sample processing, flash evaporation etc. should be separated from the main instrument room & different types of analytical instruments like GLC, AAS, IR-UV should be separated from each other by housing them in separate cabinet to minimise the problem.

Fume hoods fitted with powerful exhaust fans, should be used right at the source to remove corrosive fumes & also hot exhause coming from the flames, furnace of the instruments like Flame photometer, AAS etc.

Some instrument parts like pH electrode tips, automiser, burners of AAs, FPM etc. FID, ECD ditector of GLC etc. and measuring cells of conductivity meter turbidometer etc. develop scaling during the operation.

Such scaling may be removed first by using mild acid or organic solvents etc. & followed by repeated rinsing by distilled water.

Layer of special chemical like desicote may be given to the electrode and cells to avoid the scaling.

## PROTECTION FROM MICROORGANISMS :

The warm & humid climate enhance the growth of microorganisms, fungi & moulds, for which special attention is required.

Optical parts like lenses, mirrors etc. and chemical like standards solutions and acidic buffers have been found to be commonly affected by fungus growth.

Electronic parts like printed circuit boards, miniature contacts etc. and electrical parts like cables, insulators etc. also are likely to be attacked by the fungi & thus disturb the regular functioning of the instruments.

A thin coat of insulating varnish mixed with fungicide may be given to the sensitive electronic & electrical parts while volatile fungicidal, wrapped in cloth & kept inside the instrument compartment, can very effectively inhibit the growth.

## PROTECTION FROM WEAR/ TEAR :

It is commonly observed that the instruments, using high speed motor, get permanently damaged jut for the want of proper lubrication. Mechanical moving parts like bearings, gear slide wires, chart drive etc. require periodic cleaning & lubrication. Grease or machine oil may be used for large sized gears & bearings while special thin oil is to be used for delicate parts. Noisy potentiometer contacts can also be cured by applying few drops of potentionmeter oil.

## OTHER CONSIDERATION

a. Spot galvanometer, nopan balance & other similar parts of the instruments should be protected from mechanical shocks & vibrations.

These instruments may be housed in vibration free locations & given special plateforms. The suspensions of the galvanometer, micro-meters etc. & delicates supports of monopan balances etc. shoul dbe mechanically locked during idel conditions & specially during transit and shifting.

b. Photocells, phototues, photomultipliers etc. should not be exposed to the intense external light otherwise their sensitivities will be considerably reduced.

c. Special care is required while handling explosive gases like hydrogen, acetylene etc. These cylinders should be kept away from heat & flames. When working with flame photometer, AAS, GLC etc. simple rule to follow is that Do Not Allow the fuel gas to reach any time to the burner without the supporting gas.

Thus while igniting the flame first start the air or oxygen & then open the fuel gas. Reverse should be the order while extinguishing the flame.

d. High pressure mercury vapour UV lamps also require special care. The life of such lamps is very limited to about 100 to 150 operating hours. These lamps should not be ignited before they are cooled completely from heating caused during the previous operation, otherwise their life is reduced considerably.

Thus, it is possible to extent the useful life of the instruments by :

a. Correct planning of the analytical instrumentation laboratory.

b. By taking required minimum precautions while handling different instrument, component and parts.

c. By following proper operational procedures.

d. by providing systematic & periodic check ups & preventive maintenance.

# INDEX

## V



## W